Lithic Illustration

PREHISTORIC ARCHEOLOGY AND ECOLOGY
A series edited by Karl W. Butzer and Leslie G. Freeman

Lithic Illustration

Drawing Flaked Stone Artifacts for Publication

Lucile R. Addington

The University of Chicago Press
Chicago and London

Lucile R. Addington is illustrator of prehistory in the Department of Anthropology, the Institute for the Study of Earth and Man at Southern Methodist University. She has prepared over ten thousand drawings for a wide variety of books and articles in prehistoric archeology.

The University of Chicago Press, Chicago 60637
The University of Chicago Press, LTD., London

 Published 1986
Printed in the United States of America
95 94 93 92 91 90 89 88 87 86 54321

Library of Congress Cataloging in Publication Data

Addington, Lucile R.
Lithic illustration.

(Prehistoric archeology and ecology)
Bibliography: p.
Includes index.
1. Archaeological illustration. 2. Stone implements.
I. Title. II. Series.
CC82.3.A33 1986 930.1'028 85-8121
ISBN 0-226-00634-4
ISBN 0-226-00635-2 (pbk.)

Contents

Illustrations

Series Editors' Foreword

Stone artifacts constitute the primary research materials from most hunter-gatherer sites in both Old and New World archeology, and they commonly remain an important component in excavations of even relatively "late" sites. Consequently lithic tool typologies, notoriously difficult to master, are of central importance in archeological analysis and communication. Illustration of a representative selection of pieces from a few dozen lithic "types," picked from hundreds or thousands of excavated tools or waste pieces, is central to this communication. Critical readers of an archeological report not only must form a visual impression of the lithic assemblage but must also be able to decide for themselves whether the terminology used matches (or can be translated into) their own. Only then can they evaluate the statistical data offered. Though good photographic documentation is necessary in some situations, properly drawn artifacts are invariably more informative than photographs in illustrating a prehistoric knapper's workmanship as well as an artifact's form and diagnostic features. Consequently there is a great need for knowledgeable, professional illustrators of stone artifacts. The fact is, however, that there are far too few such illustrators, and as a result many archeologists fail to communicate their findings effectively or else opt not to illustrate their lithic artifacts at all. Particularly in "late" archeology, readers are rarely given access to the lithic finds, materials that may well convey more information about the economic activities of common folk than other archeological data categories.

Lucile Addington's *Lithic Illustration* represents a "first" in our series in that she is not a professional archeologist but a professional illustrator. But the difference is not as great as it might seem. She offers twenty years' expertise that embodies unique perspectives on scientific analysis in archeology. Her eye for retouch, snapping or truncation, grinding or cleavage surfaces, thermal fractures, and the like is rivaled by few lithic typologists. Her perspectives are often different and at times startlingly fresh and are therefore welcome as a counterpoint to the typologist's preoccupations and potential blind spots. An illustrator of Addington's caliber brings a complementary input to the larger archeological enterprise, somewhat akin to that of the geoarcheologist who can see the artifacts in the site context as sedimentary particles and can thus shed new light on the processes of accumulation. A manual of this sort is not without precedent. Indeed, the University of Chicago Press maintains in print its 1961 book *A Handbook of Biological Illustration* by Frances W. Zweifel. Like Addington's volume, this is for "artists who are not scientists and scientists who are not artists."

Lithic Illustration should interest professional or aspiring illustrators. But we believe its basic importance will appeal to a much wider audience. The unfortunate fact is that lithic typology remains a weak spot for many archeologists whose expertise lies in ceramics, faunas, architecture, and the like. This manual can serve as an indispensable guide to them and to their students by generating sensitivity to the potential information provided by lithic artifacts. Ideally it may encourage them to seek professional analysis for what may be a small and seemingly insignificant component of a site's artifacts. In effect, for those time ranges of archeology where lithic artifacts represent a minor part of the materials under study, Addington's book may well be an accessible reminder that stone artifacts encode significant information that deserves to be studied--beyond, for example, the basic recording of stone sickle blades in a Near Eastern Iron Age site or of the projectile point categories represented in a Mississippian or Anasazi megasite.

For the many archeology students who have not had the privilege of watching a flintknapper about his task, or the opportunity of taking a course or workshop on lithic typology, Addington's manual can serve as the next best thing, an insider's explanation of just what the key features and forms are and how they are used to identify lithic "types." She appreciates the fine points of typology but, at the same time, she never loses sight of the whole. Implicitly, therefore, she restores some balance to the lumper/splitter dichotomy in a subfield where the search for objectively replicable characteristics and quantifiable "types" tends to lead to overrefinement. Addington is keenly aware of the effect of raw material differences, of practical problems encountered during the actual knapping, and of subsequent weathering or reuse. This is, then, a volume that all students of lithic typology would do well to study closely and reflect upon at length.

Addington's book should also appeal to the very large group of amateur archeologists who regularly present or publish papers at the meetings and in the journals or newsletters of scores of societies. In some states these dedicated people maintain close ties with professional archeologists who often also are members of such societies and work with them. Proper artifact illustration will open important lines of communication here that would be to the advantage of all.

Finally, the effective illustration of lithic artifacts is a major problem for archeologists engaged in cultural resource management research. With inflexible deadlines to meet, such archeologists rarely have the luxury of tapping the talents of the few professional illustrators available. In such situations, Addington's manual will be just what the doctor ordered, a practical guide for artifact orientation, illustration, layout, and successful presentation that can be followed by reasonably talented people who now feel that they "can't draw a straight line."

Addington brings to this volume an enthusiasm and a talent for communicating from her long experience as a gifted artist and a devoted member of our archeological community. Her excellent renditions of lithic artifacts grace a large number of scientific papers and books, for which she has increasingly won justifiable acclaim, even among French and British specialists in lithic typology. We feel privileged to be able to include her book in the Prehistoric Archeology and Ecology series as a professional methodology, a practical manual and inspiration to illustrators (by choice or by default), and a guide to students of archeology on the significance and basic principles of lithic analysis as a critical component of our research.

Karl W. Butzer
Leslie G. Freeman

Preface

This manual provides a practical method for translating the artifact in your hand into artifact illustration whose form is internationally intelligible and ready for publication. It supplies not only the essential technical information prerequisite to drawing flaked lithic artifacts, but also the instructions for applying the information successfully. A search for other works on lithic illustration suggests this may be the only instruction manual on the subject that is written in English and covers the necessary range of practical details.

The sections dealing with the "language" of lithic illustration have a significance beyond that customarily accorded an ordinary glossary. A limited list of terms relating to lithic technology and tool typology are defined; but in addition, the conventions, symbols, and marks that constitute the substance of lithic illustration are identified and explained, and the mechanics of their use are fully presented. Drawing cannot begin until the illustrator is familiar with this information.

The personal system offered here, and exemplified by the sample drawings, follows most of the basic conventions; the methods, procedures, and style are my own. Twenty years of experience evolved the details that are my solutions to problems encountered while producing more than ten thousand artifact drawings.

The approach of this manual is purposely restrictive, especially in the limits imposed upon the variety of symbols and the kinds of marks and their placement. European illustrators, the developers and masters of this branch of art, espouse a complex assortment of symbols (many of which are rarely used) and on occasion delineate the more esoteric aspects of flintknapping, a narrow specialization unsuited to the general illustrating of large collections of artifacts. The desire to include all morphological and technological data as well as create drawings of aesthetic merit must be weighed against our need to produce enough drawings with sufficient data in intelligible form, but within time limits imposed by the real world of grants, budgets, and publication deadlines. In a university, one artist's time may be shared by several archeologists, each of whom needs hundreds of drawings every year. This pressure to produce as many drawings as possible within a scheduled time span of a few months--months that may well have been selected to phase in with expeditions and agreed upon with others, perhaps one or two years before--forces the illustrator to "get it right the first time" if possible. Most of the techniques described here were developed to speed the work without sacrificing quality.

Regarding the practicality of this system I can say, "This is how I work." This is exactly what I do to produce drawings like the ones that accompany the text. As to the scope of the material covered, everything considered extraneous to the practical business of drawing the artifacts, keeping track of the drawings, and preparing them for publication has been excluded. Historical background of the art, multiple choices of art materials and tools, discussion of other methods--all are outside the focus of this manual. The peremptory tone of the instructions may be abrasive to the knowledgeable reader, but the imperative affords a steadying support for those unacquainted with either drawing or archeology and marks the boundaries of this simplified approach. The intent is to smooth the way for those others who, while not beginners in archeology, need a little guidance in putting what they know into graphic form.

Most archeologists recognize that these specialized illustrations present data in a universal language. The ones who would like to improve the quality of what they publish by using drawings instead of photographs complain that there are too few artists with any of the background technical knowledge. These archeologists, and the artists who would try to draw artifacts for them, can find here the basic practical details peculiar to this field.

Outside this country, writings on prehistory that involve lithic material have long used artifact drawings as clarifying and complementary extensions of the text, a fact that underscores the need for prehistory students to become familiar with the meanings of the graphic devices that transmit archeological data. Students say that improved recognition of tool types and better understanding of tool manufacture are bonus results of their exposure to the fundamentals of lithic illustration.

For others, such as art students and amateur artists, this manual can make possible a beginning in this specialized branch of illustrating. However, there is one prerequisite: contact with an archeologist.

Archeologists, and the collections of stone tools in their custody, are to be found in universities, colleges, and museums. A prospective lithics illustrator is advised to investigate local institutions of this kind and make contact with an archeologist who might be willing to answer a few questions. One approach is to offer to do volunteer laboratory work, such as putting together broken pottery, inking the catalog numbers on artifacts, washing potsherds, cleaning field equipment, sorting slides and photographs, and so forth. If there is a need for such work, be dependable, accurate, careful, and unobtrusive. Such an association can provide access not only to stone tools but also to the archeologists who are the only ones who can point out to you the distinguishing attributes of tool types and explain how to read their surfaces.

With some artifacts to examine and a little coaching by an archeologist in how to recognize "bulb of percussion," "platform," "direction of blow," and such, a person who can draw, but lacks acquaintance with stone tools, can learn to follow the methods detailed here.

Still another group, those dedicated members of the societies of amateur archeologists, will find the practical information helpful. Their substantial and acknowledged contribution to American archeology could be enhanced if the artifact drawings in their publications used the conventions of lithic illustration and the layout principles discussed in this manual.

In the pages of sample drawings, you will find that the ratio of scrapers to other tool categories is top-heavy--deliberately so. Scrapers demonstrate so generously the different elements of lithic illustration--the placement of direction lines in flake scars, retouch in all its varieties from steep to deep to flat, on assorted problem contours, profiles, distal views, and so on--that offering many versions of scrapers concentrates more examples of treatments for reference.

Although the title states that the instructions herein are for drawing *flaked* stone artifacts, assemblages often include a piece or two not flaked, but pecked, ground, battered, or abraded. The illustrator is expected to draw those too. Therefore some examples of hammerstones, ground stones, and grooved stones have been added to help a beginning illustrator deal with such as assemblage mix.

Another comment about the illustrations seems advisable. Occasional inconsistencies between the artifact views advocated by the text and the ones you will see in the pages of examples are explained thus: These artifact drawings were made for assorted archeologists and not simply for an instruction manual. The *ideal* number of multiple views should always be the aim of the illustrator; the *expedient* number, sometimes specified by the archeologist and usually dictated by time/space/budget limitations, is what you may wind up with. Therefore in these pages you are face-to-face with reality from the beginning--drawings produced over a long span of time, on good days and bad, some views chosen at the discretion of the artist, some at the direction of the archeologist. Furthermore, to "doctor" or "pretty up" the drawings for their reuse here would have been a form of dishonesty; you have them as created--warts and all.

The prehistoric sites that provided the illustrated material are scattered over Egypt, Israel, Jordan, and the American Southwest. The drawings were produced for books, site reports, and articles in scientific periodicals. They are a meager sample of the thousands of my drawings published, at one time and another, in Europe, Africa, the Middle East, and the United States. Their use as examples of the drawings process was made possible by the enthusiastic generosity of the archeologists for whom they were originally drawn as part of their ongoing research. Indeed, it was their good-natured nagging that finally pressured me into writing this account of how to do what I do.

At Southern Methodist University, these are the ones most responsible for the kind of archeological illustrator I am:

Professor Fred Wendorf, who talked me out of the practice of fine art into this field, hired me to draw for him and his colleagues in Old World Prehistory, and has been the major contributor to my grants-supported job all these years;

Professor Anthony E. Marks, who willingly explains the finer points of typology, criticizes my drawings with a sure eye, and gave me the final firm push into authorship; and

Professor Joel L. Shiner, whose lucid diagrams and analysis of artifact manufacture taught me something of flintflaking, when my first year of contact with archeology was as his volunteer laboratory helper without a thought of drawing artifacts.

In the beginning, the style of my drawings was developed to satisfy the preferences of these three archeologists who have shared my drawing time.

Adapting Old World illustrating to New World artifacts was a challenge first offered by Professor Donald O. Henry, University of Tulsa, whose research includes sites in both "worlds." When nights and weekends were the only segments of my time available to work on his projects even by mail, he trusted me to meet deadlines.

In an especially thoughtful and generous gesture, the Institute for the Study of Earth and Man, Southern Methodist University, volunteered both support and crucial time for preparing the manuscript and drawings for publication. I am grateful to Professor James E. Brooks, President and Director.

The Department of Anthropology, Southern Methodist University, provided valuable help with manuscript preparation. My thanks to its chairman, Professor Ronald K. Wetherington, the faculty, the staff, and both current and former students for their offers of assistance and their friendly interest in the progress of this book from its inception.

Periodic lectures to graduate students, at the request of Professor Fred Wendorf, forced me to organize my thoughts with notes that eventually became the springboard for creating this manual. My thanks go to him and to his wife, Professor Angela E. Close, for proffered permission to use any drawings I had made for their many books. Also, my thanks to their colleague Professor Romuald Schild, Institute for the History of Material Culture, Polish Academy of Science.

I am particularly grateful to Professor Anthony E. Marks, whose constructive criticism, suggestions, and patient reading of manuscript sections during the writing process sought to steer me around some technical gaffes; any that remain are my responsibility. It was largely his supportive interest, as well as his unrestricted permission to use drawings I had made for him, that encouraged me to broaden my early conception of this book from sparsely illustrated instruction manual to a more widely useful reference book of sample artifact illustrations.

I appreciate the warmly generous response of Professor Donald O. Henry, University of Tulsa, with his permission to use my drawings that illustrate his research in both this country and the Middle East.

It was Professor James L. Phillips, University of Illinois at Chicago, who mentioned the existence of my manuscript to Karl W. Butzer. Professor Butzer's immediate recognition of the need for such a book finally put it in the hands of the University of Chicago Press. I am grateful to all those who helped get it into print.

Special thanks are due John F. S. Phinney, Director of the Library of the Institute for the Study of Earth and Man, who gave me unlimited access to his professional expertise and brought bibliographic order to the references. Equally important to me has been his friendly service as sounding board in all circumstances.

My thanks also to Laurie Evans for reminders, suggestions, and the typing of my left-handed scrawls, during early months, when an injured right hand made both typing and drawing impossible for a while. Her later voluntary sacrifice of vacation time to assist with the layouts was crucial to their preparation.

Susan Meyn's expertise in word processing was a major contribution to final form of the book.

My greatest debt is to my husband, who encouraged my efforts, read every chapter I thrust into his hands, yet managed to restrain his own professional journalistic talents so completely that never once did he take any part in the writing. Instead, for me to have the time to produce this manual, he shouldered some of my routine chores without complaint.

Sheer numbers--nearly a hundred chapters, sections, books, and articles illustrated for more than thirty people--preclude listing many to whom I owe much; but I appreciate the associations that enlarged my experience with a variety of artifacts and approaches to archeology.

Always a source of inspiration toward excellence, the extraordinary drawings by the great French illustrator Pierre Laurent, University of Bordeaux, have been an authoritative reference on many points through the years. His generous answers to my technical questions have become a continuing correspondence that lessens my sense of isolation from the mainstream of archeological illustrating.

I am indeed grateful to the classes of graduate students of prehistory who let me bully them into following my instructions and thereby prove that my way of doing things can work for others too.

Certain spellings and the bibliographic style are those specified by the publisher.

Introduction

Lithic illustration has one foot in archeology, the other in art, and its mind on the future published page. A prospective illustrator needs at least a nodding acquaintance with all pertinent matter before drawing can begin. Many of the decisions made, and actions performed, during the drawing process are dictated by the requirements of later procedures.

The teaching of this branch of illustration has been largely oral in tradition, and the learning of it supplemented by reference to published drawings executed by other artists. Communication of the subject is strongly person-to-person, with its descriptions and instructions couched in informal language that nonetheless incorporates technical terms. In the case of lithic illustration, it takes "a thousand words" and more, to tell how to draw the picture. Cutting this spate of conversation to a minimum in the chapters on the actual drawing process makes their sequential steps a tidy ordering of what to do next. The necessary definitions and explanations are grouped in other chapters whose contents, it must be emphasized, are essential to following the instructions.

The subject matter falls naturally into broad divisions that deal with technical information, directions for the drawing of the artifacts, and preparation of the drawings for publication. Although these divisions and their segments must perforce go single file, their order does not necessarily imply relative importance or sequence of use.

A portion of the information, the specialized terminology, is in the form of simplified definitions that may impress the knowledgeable as indeed simplistic; but they are meant to aid the totally uninitiated. More exact and exhaustive definitions of tool types can be found in Brezillon (1968), Bordes (1961), de Sonneville-Bordes and Perrot (1954, 1955, 1956a,b), Marks ed. (1976), Laplace-Jauretche (1957), Movius et al. (1968), and Tixier (1963).

Not to use the terms that label tool "types" and core "types" would make communication impossible--between archeologists and illustrators as well as between this manual and those who consult it. The limited glossary in Appendix 1, coupled with the above sources of tool type definitions given above, should make the instructions intelligible. Although some familiarity with the names of tool types is desirable, to offer more than is presented here would range beyond what is essential. Moreover, "typology" and "typologies" are complex subjects; discussing them is the province of archeologists. In practice, it is preferable that the beginning illustrator be exposed to the personal typologies of the archeologists for whom the illustrating is done. As interest in typology is fostered by the archeologist's explanations and the experience of drawing, the subject can be pursued independently.

Working for two or three archeologists, an illustrator's acquaintance with tool types is inevitably limited to the assemblages from their special areas of interest and research. To fill the gaps in such a fragmented experience as well as to clarify aspects of the tool-making process, I suggest that the illustrator audit a course dealing with typology and the classifying of tools. It will benefit the archeologist, in the long run, to have an illustrator with a more knowledgeable approach to the work--one capable, finally, of drawing with little or no supervision during much of a project.

When an illustrator has become experienced in the practice of drawing stone tools--has learned to "read" an artifact's surfaces, gained some insight into the archeologist's approach to typology, and knows the range of tool types likely to appear in that individual's collections--then the following advantageous and efficient option for scheduling illustration time becomes possible and workable. The collection of artifacts to be drawn, together with their pertinent descriptive field notes, can be left with the illustrator while the archeologist is, perhaps, on the other side of the world, engaged in another field season's work. For this plan to operate successfully, the archeologist should first classify the artifacts for illustration in boxes labeled according to tool or core type and flag their descriptions in the field notes. The illustrator's responsibility, then, is to make the necessary drawings of each artifact, based on the data in the field notes concerning its attributes and special features. There may be a few problem pieces that require authoritative decisions, but they can be dealt with speedily when the archeologist returns if a small memo is taped temporarily beside the pertinent field note and a similar one, with a large question mark, is placed in the box with the artifact.

Most archeologists avoid assigning functions to tool types, parrying the layman's natural question about the purpose of a specific tool with "We can't be sure what it was used for." This answer, in the face of such labels as "scraper" and "perforator," seems to carry scientific caution beyond reason. When pressed for an explanation, they may say that the labels are based on the broad assumption that form follows function. For example, to punch a small hole in a hide, a large flat piece of stone would not be morphologically suitable for the job.

In recent years there has been a burgeoning interest in studying the signs of use and wear on stone tools, partly in an effort to explore their function. However, until the results of further research provide firm evidence, archeologists will continue to answer in their cautious way.

To pursue this subject of function in more detail, see Cahen, Keeley, and Van Noten (1979) on microwear patterns, Keeley (1980) on experiments to determine stone tool uses, and Odell and Odell-Vereecken (1980) on low-power microscopic techniques for use-wear analysis.

Information on orienting general tool types for drawing will serve reference purposes best in stringently condensed form. Still other aspects of drawing and of layout composition are cast in discourse informal enough to allow the unavoidable mixture of commands, cautions, and instructions that must encompass the minutiae.

If there were a way to present all aspects of artifact illustration at once, the eager and the impatient could open the book, seize a pencil, and execute "step one." However, that satisfying move to get on with it must be postponed until an assortment of information has been examined. So interlocking are the various aspects, that what is past and what is in the future have a say in what is done at the moment.

It has been a primary concern, in creating this instruction manual, to provide a generous array of artifact drawings not only to serve demonstration purposes in connection with the text, but also to do reference service as models long after the initial learning period has become routine practice of the art. With that future use in mind, the sample drawings were selected, from the thousands available, to show how to handle certain problem contours in different views, and how to manipulate the marks and place the symbols discussed in the chapter on that subject along with treatments for specific features and surfaces. Any of the drawings, as a matter of course, would embody some of the conventions, but most were chosen to demonstrate several items of information. For example, the drawings of a single artifact may offer not only outline, direction lines, cortex, steep retouch, and a distal view, but older patinated flake scars and fire-popped pits as well (fig. 11, *d*). The graphic devices in use on a variety of pieces, originally drawn for archeologists to publish, constitute a display more broadly helpful to the beginning illustrator than a few choice drawings produced solely for a book about drawing.

Not every drawing in the figures is referenced within the text. Many of them are present with the expectation that someday they will match up with problem pieces and be helpful. There is another, and even more important, reason for having so many drawings here: the more individual tools and cores in one's experience, the more one absorbs about their range of variations and their patterns of manufacture.

Two or three endscrapers--or burins, or denticulates--cannot begin to educate the eye or hint at the variety the illustrator will encounter. Becoming familiar with these drawings can help make up for a possibly limited learning period. Also, looking at these drawings again and again will tend to increase confidence in following the instructions and should influence the style and the look of the resulting work.

The range of tool types is general, not exhaustive--simply samples from both Old and New World collections. However, the list of references includes sources where additional types and their variations are illustrated by hundreds of drawings.

Exactly what constitutes a good layout, or makes a bad one bad, is hard to define, as are most subjective judgments concerning proportion and balance. To focus attention on the premise that "good" layouts help and "bad" ones hurt, the chapter on layout composition offers more than airy theorizing. It points out some moves that will skirt the most blatant faults and encourages tinkering with a layout arrangement to arrive at the best effect possible. In the real world, not every layout can be superbly balanced--a stately procession of diverse shapes that obligingly accommodate each other's ins and outs and sizes; you work with what you have. That aspect is explored further in the chapter on specific layout problems, with suggestions for handling some typical awkward situations.

Photographic reduction of drawings for publication is a sensitive subject from the artist's side. Years of recording the evidences of prehistoric man's labor--data meticulously transcribed from artifact to artifact illustration, piece after piece--all, shriveled and blackened at the rolling of a press! That is the artist's nightmare and occasionally the artist's experience. In some cases defensive drawing can try to allow for the effects of reduction and help minimize the possible damage, a maneuvre discussed in the chapter about reduction. But still, your drawings are vulnerable to poor quality paper and antiquated presses. To know this and still maintain a high personal standard in your lithic illustrating is the only course to follow.

From the moment of discovery, each artifact has an assigned identity, and that precise individual designation must never be separated from it. The illustrator, no less than the other members of the archeology staff who handle artifacts, has a basic responsibility to keep track of these identities; but in addition, the artifact drawings and their layouts must be recorded in a way that prevents mix-ups and also facilitates retrieval from storage. Since that duty cannot be isolated from artifact drawing, suggestions on record keeping are included here.

Reaching that seemingly distant goal, publication of the archeological findings, involves coordinated work by specialists. It is not surprising that one specialist can have a myopic view of another specialist's problems and concerns, if indeed there is any awareness at all. Although tangential to the drawing of artifacts, the working relationship of two of these specialists--the archeologist and the illustrator--does have some bearing on how smoothly the progress toward publication proceeds. Bringing forward a part of the practical functioning of the relationship may prove helpful.

In the natural course of practicing lithic illustration over a period of years, a variety of small problems have time to arise, and ways of handling them evolve. Useful as the solutions are, their presence would tend to disrupt continuity if they were tucked in wherever they might fit the subject under discussion. Therefore all such trivia have been collected, like household hints in the back of a cookbook, toward the end of the manual (Appendix 3). To leave them out altogether would diminish the principle of learning from another's experience.

The real history of archeological illustrating is left for someone else to explore; but those with strong interest in the subject can find something about its early days in Stuart Piggott's *Antiquity Depicted* (1979). Meanwhile, more appropriate here, some speculating about the origins of seemingly arbitrary procedures may make the advocacy of their practice more palatable.

Archeologists frequently do their excavating where a succession of others of their kind have worked for scores of years, or they may deal with time frames or cultures already investigated elsewhere. To place their research in the scheme of all that has gone before, in generations of work in the same areas, whether geographical, cultural, or chronological, they must know what discoveries resulted and then were described in the literature of archeology.

The writings of those forerunners had accompanying illustrations that are still examined and referenced by today's archeologists. Specific tool types were drawn in certain poses. Symbols were created, assigned meanings, and added as needed. In the beginning, the logic of the orientations and the symbols was probably that of the archeologist. Ultimately, those decisions of the time became conventions by imitation, by reiteration, and by their familiarity within the archeology community. As one generation's research linked to another's, so the conventions of the illustrations depicting their findings became general usage and, to some extent, codified. Illustrators, too, must have exerted an influence.

Shifting emphasis in research, and the more refined studies now possible with the instruments of advanced technology, may well call for invention of graphic devices and symbols beyond what has been adequate for current use. However, the important thing is to draw in the manner that is familiar to the rest of the archeological world and to be very sure that an invented symbol is truly necessary. For an illustrator to change the "language," in toto, out of caprice or arrogance or laziness, can only make the drawings incomprehensible and result in misinterpretation of the data.

If archeology interests you beyond mere ink lines on paper--and surely it must!--look through *A Hundred and Fifty Years of Archaeology* by Glyn Daniel (1976). Admittedly remote in its relationship to the drawing of stone tools, it is nevertheless illuminating and helpful once an interest in archeology is aroused. Its condensed chronology of archeological events gives the perspective of a time frame to such things as discoveries, research in different areas of the world, names of the greats in archeology, their publications, and the eras spanned by their careers. It is a valuable source of references for further reading.

For a detective story approach to archeology, if one is only newly attracted to the broad field, *Gods, Graves, and Scholars* by C. W. Ceram (1951) can be an engrossing and entertaining introduction. It unravels the mysteries of a few of the most famous events in archeology by tracing the clues that led to major discoveries, as well as supplying enough biographical material about the discoverers themselves to transform famous names into real people.

Illustrating the stone tools unearthed by prehistorians can have a unique fascination, particularly for those who will admit to an idealistic desire to help fill in the blank spaces in modern man's knowledge of early man. There is a superficial sameness to broad classes of tools, and yet each piece is different from the one just drawn and the one yet to follow. Drawing the same types in quantity need not become repetitious labor if you will make an effort to understand what you see and draw. Focus "small" in the intensity of your observation and sort out the evidence that can be read on the artifact in your hand. Such concentration in the performance of stone tool illustration is often punctuated by intriguing little discoveries concerning patterns of manufacture--signs that a flaw in the raw material caused problems for that toolmaker, indications that he tried to salvage a botched job or convert a tool to a different purpose. At times, clues may suggest that not all the work on a piece was done by the same pair of hands--more imaginative guesswork than science! But any musings that trace these pieces of rock back to their shaping by human agency are entertaining accompaniment to the solitary work of illustrating stone tools.

"I can't draw a straight line" is not an acceptable excuse. Doggedly refuse to let any artifact be "too difficult" for you to draw. The chapter on views and the frameworks for drawing them details the use of guide lines with which to conquer even the most complex retouch, no matter what secondary views of the piece are called for.

Aided by pairs of tiny feature markers, discussed in the chapter on making the pencil drawing, you can establish the location of enough landmarks within a large area of flake scars--say, for example, those on a projectile point or a biface--to make drawing the rest readily accomplished. The method eliminates much of the frustration of erase and try again.

The basic conventions of depicting flaked stone artifacts are not to be flouted, especially the most ironclad of all: light must be from the upper left, at a forty-five-degree angle, illuminating the artifact in all views; and another: the profile view should be drawn beside the edge whose data it displays.

I advocate certain drawing instruments and materials because they work best for this system of illustrating. The reasons for their selection, such as the nonsmearing, point-retaining qualities of 5H drawing pencils, are explained in the chapters on penciling and inking. If you want to model the style and effect of your drawings after the examples supplied in this manual, use the same kind of drawing paper (Strathmore single-ply, medium surface) and the same kind of pen (Hunt's no. 104). Tracing paper and a reservoir pen will not produce drawings with the look of these.

The more drawing you do, the better at it you will become. Soon each new collection of artifacts from the field is anticipated with relish, and a complex artifact is a challenge that gives satisfaction in the successful illustrating of it. Decide that you will be attentive to details, self-disciplined in keeping the quality of your work high, and determined to draw any piece that comes your way. The necessary instructions are here, and sources for further references are listed; the rest is up to you. Familiarize yourself with the basic information and get on with the drawing.

Part 1

The Substance of Lithic Artifact Illustration

1 The Marks and Symbols

Certain code-like graphic devices--marks and symbols--are the means by which you will make the artifact in your hand take form on the paper. The discussion here includes only those devices that experience has proved most often needed in this method of lithic illustration. Setting a limit upon their variety, and describing exactly how to use the marks and symbols advocated, simplifies style and places application of the fundamentals within reach of modest drawing skills.

What the marks and symbols are, what they mean, and how to make them function is a body of information usable only if you know how to obtain the data they are meant to express in the illustrations. "Reading" the surface of the artifact itself provides that data.

Two items of equipment are essential for this examination of the artifact: a small incandescent high-intensity lamp, placed close by at your left, and a magnifying glass mounted on a base. Fluorescent light is unsuitable; it wipes out the shadows that are necessary not only for reading the surfaces, but for determining where to attach certain lines in the drawing. The magnifying glass will be in constant use to make a multitude of decisions with more certainty and speed.

The placement of the lamp provides for "light from the upper left," meaning that all artifacts must be illuminated for drawing by lighting that strikes them from the upper left, at an angle of about forty-five degrees. This is a convention so universally sanctioned that it is the most inflexible rule in lithic illustration. This standard angle of lighting casts all shadows in the same direction in every drawing and unifies any group of such illustrations on the printed page.

A magnifying glass three or four inches across is ideal. The power of an ordinary "Sherlock Holmes" type of reading glass is sufficient; but the magnifier must be mounted on a base that will support it at a convenient height and angle near the lamp, freeing the hands to hold and manipulate an artifact in the correct angle of light as the magnifier is used.

For the special needs of illustration, here are some pointers on how to look at a stone tool as well as how to recognize and interpret what is there.

First, find the tool's ventral surface. It will possess either the bulb of percussion or, if it is missing, surface evidence of its original position. Most of the subsequent readings and determinations are based on the location of the bulb in relation to other areas of the artifact. Keep the bulb end at the bottom, or at least remain aware of where it is, while you turn the piece to examine other surfaces and features.

This analytical inspection is a way of becoming acquainted with the piece, an essential preparation for drawing it. Think in terms of such questions as these few examples:

Where are the ventral surface and the bulb of percussion? (That will establish the location of the "back" of the piece and the "bottom" or proximal end, data upon which to base other decisions.)

Is this small flaking to be considered retouch? Backing? Truncation? (If it is steep enough to be considered backing, it probably requires a profile view. If it is not only steep but obviously cuts off the distal or proximal portion of the piece, it is a truncation and calls for adding a truncation view. If the truncation is oblique, its view may have to be drawn in the oblique position.)

Is this scattering of small scars at the edge retouch? (That the scars are scattered excludes "retouch." If the color and texture do not match the rest of the piece and the scars look freshly chipped, it is recent accidental damage; omit it from your drawing. If the color and texture match the rest of the piece, it may be "use retouch," and its exclusion from the drawing follows the policy of the archeologist, although minimal "use retouch" is commonly omitted.)

In the same way, sort out the pattern of manufacture as this example suggests:

Which of these two scars came first, and where did they originate? (If scar A removed the impact area of scar B, then B was produced first. Further, if this scar A is on the edge of the piece but its point of impact is not present, it could not have originated there. Therefore both A and B had to be produced before the piece was detached from the core.)

The raw material of a tool may have a texture that is hard to "read." Manipulating the artifact in relation to the light source can help determine the directions of blows. Try holding the tool near the edge of the lampshade so the amount of light on the piece is drastically reduced and crosses the surface of the scars at very low sunset angle. Slowly tilt and swivel the piece while watching for movement of shadows and highlights whose curved path of travel is the clue for the curvature and placement of direction lines in the drawing.

For another way to determine direction of blow, look closely at the lateral boundaries of a flake scar for minute feathering and little runners or ladders (fig. 1) whose one free leg will dangle inside the scar to which it belongs, frequently pointing in the general direction of the impact area of the scar. These tiny runners, often exhibiting miniature hinge fractures at their distal ends, are useful clues not only to directions of blows, but also to sequence of blows, in that the flake scar to which they belong must, of course, overlie the adjoining scar; they would have been removed if the neighboring scar were the product of a subsequent blow and therefore "on top."

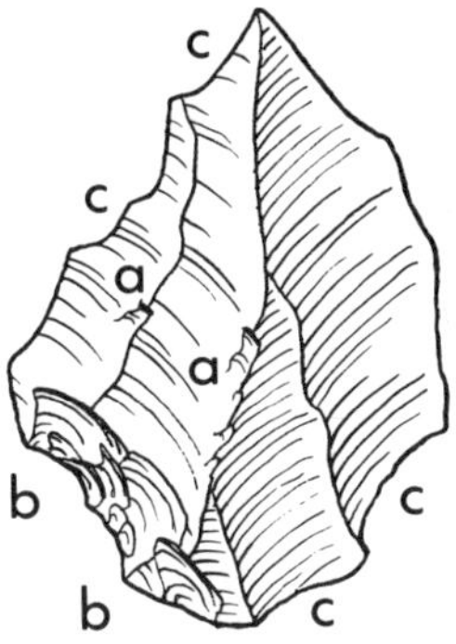

Figure 1. Overlaping of flake scars. *a*: runners, ladders, feathering; *b*: flake scars with point of impact on the edge of the piece; *c*: flake scars that originated on the core before the piece was detached.

On a core, the first feature to look for is a platform; then, on an adjoining face, look for the flake scars that resulted from removals of material when the platform was struck. A platform, usually scar surface but occasionally cortex, looks incomplete or "bitten into" along the edge where it meets the adjacent face's points of impact. Look for any other platforms that may be present, and follow the instructions for orienting and drawing different types of cores.

Cortex, the natural exterior rind of a rock, was subjected to every altering and destructive agency in its environment for a longer time than were the flaked surfaces of the interior. Consequently, the cortex on an artifact tends to chip or crumble where it meets a flaked surface at a distinct angle. Be suspicious of what at first glance resembles retouch scars in the edge of the cortex. Judge severely; let your drawing include it only if the flaking appears regular and purposeful.

Here are the marks and symbols to use. Appended to the first list is another, amplifying its code aspect by detailing the treatments for additional specific artifact features.

Outline of the Artifact

Draw the outline of a piece with an effort to reproduce its characteristics, which can range from crisply edged with minute sharp projections (fig. 15, *f*) to smoothly rounded (fig. 2, fig. 40, *f*). Both the raw material from which the piece was made and the forces that acted upon it after manufacture determine the condition of the edge and consequently are a part of that artifact's individual data.

Outline of the Flake Scars

Draw each scar as an entity. If a scar is whole, draw its outline that way; if it is the remnant of a scar, with part missing, its outline will end against the adjacent, and subsequent, scar's outline (fig. 1, fig. 44, *c*). Keep in mind that intentional blows or pressure produced flake scars singly, and repeated applications of the force created overlapping rows, or layers, or tiers of scars. Therefore parts of neighboring scars were usually obliterated as successive removals continued during manufacture. Since this sequence of removals is part of the data related to the artifact's pattern of manufacture, draw the scar outlines carefully to reflect this information. It is a common fault for flake scars to be drawn as if the ridges were the meshes of a net instead of the outlines of individual scars.

For pieces so weathered or sandblasted that the scar ridges are rounded off and, in some places, difficult to distinguish, draw the more prominent ridges with an irregularly broken line (fig. 2). Pencil the outlines of the other blurred scar shapes but do not ink their outlines: ink only the direction lines within the scars and erase the penciled outlines. This treatment produces a blurred effect in the illustration and is truer to the condition of the artifact than sharp delineation would be.

Direction Lines

Direction lines, in artifact illustration, are a series of curved and nested lines whose placement indicates direction of blow or pressure. They are drawn to arch across the path of that force in each area of

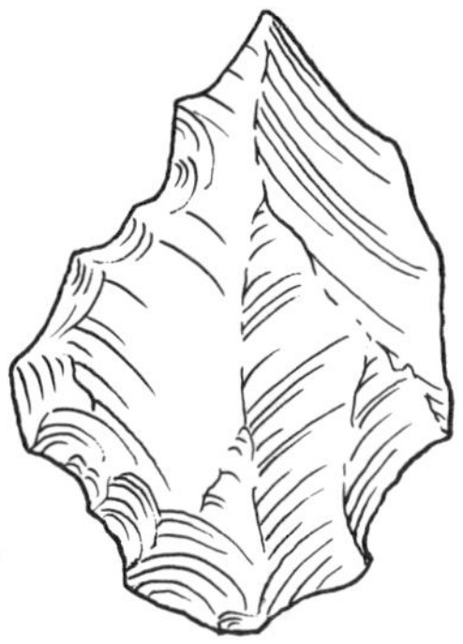

Figure 2. Severely weathered flake scars.

artifact surface exposed by the flaking process, as in figure 11, *f* and figure 25, *b*, for example.

Direction lines are stylized representations of the fine ripples that can be seen spanning both surfaces created by detachment of a flake of stone: across the ventral (bulbar) surface of the flake that was removed and, in complementary form, across the scar produced by this removal. Caused by the force of detachment, these minute undulations in stone are analogous to the concentric ripples produced when an object is dropped in water. They too curve around, and radiate from, the point of impact.

In illustrating flaked lithic material, direction lines are the most used convention of all the code of marks. They are the main graphic device with which to incorporate data about the manufacture of the piece, while at the same time their manipulation by the illustrator gives form and solidity to what otherwise would be a flat diagram.

"Direction of blow" is the information they supply, but direction lines also provide a means of shading the drawings. Lights and darks, and gradations between, are produced by the spacing of the fine black lines on the white paper; the more lines, and the closer together they are, the darker the effect. For this reason, direction lines should be varied in length and spacing to give the tones of light and dark dictated by the conformation of the surface and its position on the artifact. (This underscores the importance of correct lighting of the piece.) Do not extend direction lines completely across an area. On a flake scar they are attached, at one end only, to the most shadowed side of the scar (fig. 25, *f*, left view); on a bulbar surface, they are attached to opposite sides of the surface and allowed to interfinger as the series leaves the area of the bulb itself and proceeds toward the distal end (fig. 25, *f*, right view). Although not adhered to rigidly by European illustrators, these dogmatic restrictions can be justified as a simplification of style that makes explicit the distinction between "flake scar surface" and "bulbar (ventral) surface." Furthermore, no great artistic ability is required either to apply the principles of direction lines in this restricted manner or to draw them so that their meaning is unmistakable. Attached to one side of an area, they mean "flake scar"; attached to two sides of an area, they mean "bulbar surface."

What goes into the outlined areas of the illustration concerning direction of blow is to be found on the corresponding surfaces of the artifact itself; *where* to attach direction lines in the drawing, once the direction of blow is discovered, is determined by where the shadows are when light from the lamp strikes the artifact from the upper left.

In general, the upper left areas of the drawing will be lightest in tone; the lower right areas, away from the light source, will be the darkest. However, that oversimplification is tempered by the form and surface irregularities of the individual artifact. Hold the piece so that the illumination strikes it from the correct angle; then the lights and darks and middle tones are in their proper places to show where to attach, and how to distribute, direction lines.

A brightly lighted flake scar needs only a few short, widely spaced direction lines, merely to indicate the direction of blow without darkening the area too much. A scar in a heavily shadowed area needs more direction lines drawn closer together and extending farther across the scar.

To show pronounced ripples in a flake scar, repeated groupings of some of the longer direction lines produce the effect (fig. 15, *e*).

In all cases, vary the lengths of the direction lines as they are drawn into a single area, one after another. To draw all the long ones, then go back and put the short ones between, is never successful. Graduate the lengths as they are drawn into a scar so that the tones of shading they produce will be smoothly graded.

When you have drawn all the flake scar outlines and are ready to fill in the direction lines, go about it in this way. Look at a scar while the artifact is held in the correct angle of light. Decide where that scar is darkest, lightest, rippled, and so on. Then, ideally, draw the whole series of its direction lines across it, one after another, in a single operation, from point of origination to its distal end. This treatment of an individual scar in a continuous application of its direction lines is especially important when inking the drawing, because the ink dries on the pen rather quickly if there are repeated pauses to check the artifact. Wiping the pen, reloading with ink, and returning to the same scar results in lines that do not match in weight or grade evenly in curvature.

A flake scar displays at least a slight hollow nearest the point of impact, and indeed some scars are deeply hollowed in this area that produced the bulb of percussion on the flake that was removed (fig. 12, *g*). Begin drawing the direction lines here. Attach them to the most shadowed side of the scar as determined by the correct angle of lighting. Begin at the point of impact, where the direction lines should curl most tightly, and rapidly reduce the amount of curvature with each successive arc of the series as distance from the point of impact increases. At the same time, vary the spacing to produce the necessary shadows and textures.

If the point of impact is present on the artifact, the tight curvature of the first few direction lines indicates that fact. If the scar has had its point of impact removed by subsequent scars, arcs of much less curvature will mean it is a distal remnant of a scar (fig. 13,

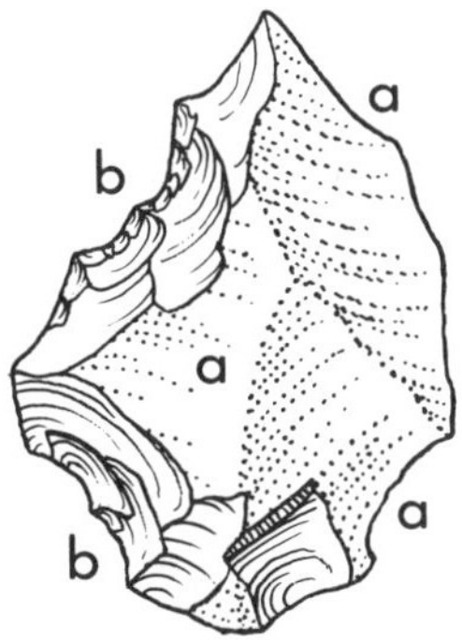

Figure 3. Two ages of work on the same piece. *a*: heavily patinated older flake scars; *b*: younger scars.

i, major scar). If a scar appears to have originated on the core, before the piece was detached, direction lines placed at the edge of the piece in that scar should be drawn with very little curvature, an indication that the scar's point of origination was not on the edge itself (fig. 15, *e*, major scars).

On the ventral (bulbar) surface, direction lines are attached at one end to both lateral margins, their free ends interfingering deeply over the length of the surface (fig. 26, *c*; fig. 33, *d*). However, begin at each side of the bulb of percussion where it joins the platform at the point of impct and start the series with a few short arcs that curve as if to encircle the bulb itself, gradually increasing their length and decreasing their curvature as the series leaves the area of the bulb and continues toward the distal end of the surface with lines that are almost straight.

Define the bulbar cone by two small lines of unequal length, like an irregular letter V with its bottom touching the point of impact at the platform and the bulb in its fork. If the proximal end of the ventral surface is missing, because of truncation, snap, retouch, or a break (fig. 27, *f*), no mark for the bulbar cone is needed.

For the occasional artifact with two ages of worked surfaces, as evidenced by marked difference in degree of patination, follow the rules for direction lines but use arcs of evenly spaced dots, instead of lines, to show direction of blow in the heavily patinated scars (fig. 3; fig. 18). This is not to be confused with cortex, whose stippling is deliberately random and has quite another meaning.

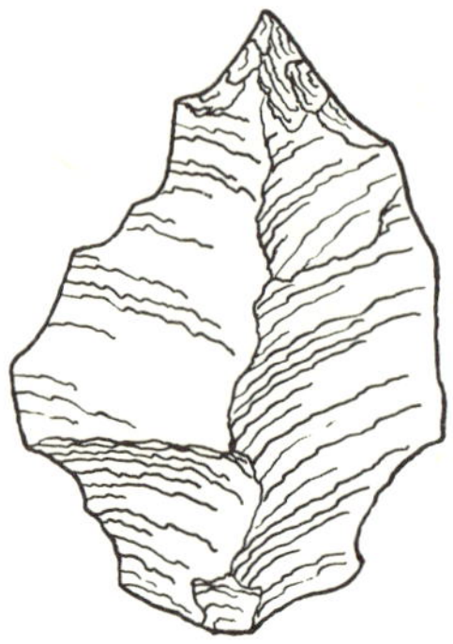

Figure 4. Direction lines for rough-fracturing material.

Some types of quartz and other rough-fracturing material (fig. 4; fig. 30, *d*) can be suggested by direction lines drawn in a shaky manner.

Stippling

Stippling is the graphic device of rapid, random dotting with ink on a specific portion of an artifact drawing. It identifies and, by the manner of its application, distinguishes cortex, ground and polished surfaces, and areas of sickle sheen.

In lithic illustration, stippling can enhance the three-dimensional quality of a drawing if it is executed with some care. It takes more time than the treatment of flaked surfaces, but the illustration of solid form is dependent on the quality of the stippling, especially if an artifact is largely covered by cortex.

To draw cortex (fig. 50), repeatedly meander the dotting process over the area in order to avoid creating any recognizable patterns. Study the lights and shadows of the cortex surface on the piece and reproduce the tones by stippling little for light areas and more for dark.

If the cortex is smooth, keep the dotting rather fine and uniform, with gradual changes of tone. If the cortex is rough and coarse, stipple with considerable vigor in some areas, according to the character of the surface features.

One common problem is an unfortunate aspect of coarse dots scattered over an area on which they appear to float in a flat plane. The solution is not only more dots, applied with greater care, but also dots stippled right into the line that is the boundary of the cortex area. this will eliminate the strip of white between stippling and outline That gives the effect of floating dots.

Ground or polished surfaces (fig. 54, fig. 55), such as those that occur on grinding stones and celts, must be stippled with special attention paid to the subtleties of lighting on curved surfaces. Use the finest stippling on the smoothest areas and coarser dots on other areas that may be pecked or pitted, or simply cortex.

Illustrations of artifacts with sickle sheen (fig. 53) are more satisfactory if you draw "repeat outlines" in which to display the sheen than if you try to depict areas of sheen mixed in with direction lines on the primary views of the piece.

First draw the artifact in the usual way as its tool type dictates; then, beside the edge with sheen, make a full-length repeat outline of that edge only and stipple in the sickle sheen areas, leaving out of the second drawing everything else, such as flake scar ridges, direction lines, and so forth.

If there is no inverse retouch or other features that would require a finished drawing of the ventral surface, the back view of the piece can be a complete outline of the artifact with only the platform outlined, the bulbar cone mark indicated, and the ventral occurrence of sheen stippled in. If a complete ventral drawing is necessary, follow the procedure of a full-length repeat outline of the pertinent edge and stipple the ventral distribution of sickle sheen.

All of the foregoing discussion of stippling and its uses applies solely to the inking process. For the pencil stage of the drawing, a loose circular scribble is enough to remind the illustrator that an area is cortex, and some extra scribble will locate shadows and such features as hollows, folds, or lumps. It is a waste of time to reproduce exact shading in pencil that will have to be erased and that, indeed, interferes with judging the tones being stippled in ink.

On grinding stones and celts, a little pencil shading will indicate areas of shadow to be stippled, and a faint pencil line is helpful to show where to end fine stippling and begin coarse stippling.

Use a light pencil line to mark the limits of sickle sheen and some pencil shading to locate the brightest portion; then finely stipple with ink, applying fewer dots where sheen is fainter and more dots where it is brightest. If retouch occurred in the area subsequent to production of the sheen, and therefore removing portions of it, do not outline the retouched surfaces in the repeat view, whose only purpose is to show distribution of sickle sheen. Stipple carefully to leave those areas empty. An example of this would be a piece that was resharpened after considerable use.

The quality of the finished drawing is improved by the technique of continually moving to different parts of the entire area as you apply the stippling, returning again and again to add more, so that all portions are brought along at the same stage of finish and none is worked from beginning to final stage by itself. This is especially important in the case of multiple views of a piece with much area to be stippled (fig. 60). Keep all views approaching finished stage at the same rate by stippling some on each view, repeatedly. When the lights and darks have been located on the drawing by tentative preliminary stippling for special features such as knobs and furrows, erase the pencil scribbles so that the ink stippling can proceed uninfluenced in tone by underlying pencil shading.

As always in the matter of light and shade, the artifact itself, lighted from the correct angle, shows where the shadows are and their relative intensity. Squinting at the piece with half-closed eyes will help you check your drawing as you gradually develop the form with stippling.

Dashed Lines

Extend several small uniform dashes from each end of a break to show that part of the artifact is missing and how one assumes the missing piece must have continued the outline (fig. 5; fig. 43, *i*). Consider the general trend of the unbroken portion of the edge and the thickness of the piece at the break to make logical placement of the few dashes. Avoid the temptation to reconstruct the entire outline of the missing portion; that assumes too much.

Dots

Used in a row, in various ways that have quite different meanings, dots are not to be confused with the random dotting (stippling) of cortex, for example.

For the outlines of older, patinated flake scars (fig. 3; fig. 18), use a line made up of fine, evenly spaced dots to distinguish those scars from the younger work on the same piece. Furthermore, make the direction lines, within the patinated scars, dotted curves instead of

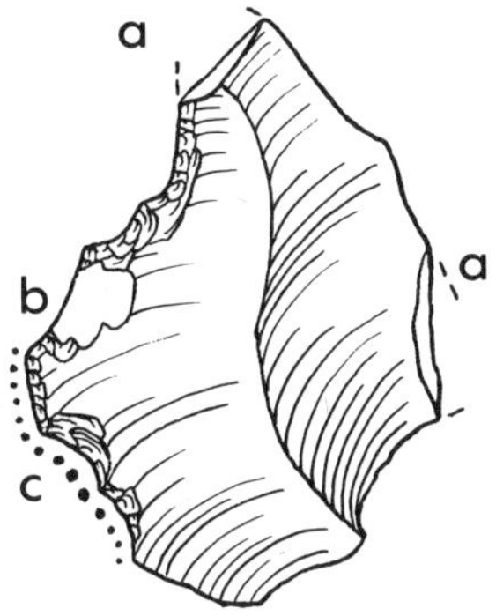

Figure 5. Breaks and abrasion. *a*: dashed lines for a break; *b*: "bag-broken" area; *c*: row of graduated dots to indicate abrasion.

solid curved lines like those within the newer scars. Where the more recent scars meet the patinated area, note that the outlines of the newer ones, drawn with a solid line, prevail because they are subsequent to the older scars.

Abrasion that has dulled or blunted an edge can be indicated by a row of closely set single dots outside the primary view, near but not touching, alongside the affected portion of the edge to show the extent of the abrasion. The dots may be alike or may be varied slightly in size to suggest degree of dulling: smallest ones for places with minimal abrasion, dots of increasing sizes where degree of abrasion becomes greater (fig. 5; fig. 13, *c*).

Another conventional symbol for grinding or abrasion on an edge is simply one dot placed beside the drawing at each end of the affected portion, with a reference in the descriptive matter such as "grinding between dots." This provides a less cluttered effect, but it gives no details on degree of abrasion.

Arrows

Simple straight arrows (fig. 6; fig. 32) are reserved for use solely with drawings of a general tool type called "burins." Place an arrow outside the drawing of the piece, near but not touching it, to indicate the direction of each burin blow that removed a spall and affects the present shape of the tip. Tentative, small flake scars in the area of the burin bit do not rate arrows, nor do distant remnants of burin blow scars whose points of origination have been obliterated by subsequent burin scars.

Illustrations of burins, whatever the variety, usually need multiple views of each piece. What does not show in the front or the profile views may need a back view to transmit all the data. Transverse, busquoid, and bec burins often require distal or proximal views.

In dorsal and ventral views, a qualifying burin scar is entitled to an arrow if the scar, visible or not in that particular view, is immediately adjacent either to the main surface or to another burin scar that is. Dihedral burins (fig. 32) always get their characteristic convergent slanting arrows with both dorsal and ventral views, even though one of the burin facets may not be visible. In profile views of dihedral burins, only burin scars that are visible in that position get arrows; the out-of-sight dihedral scar does not.

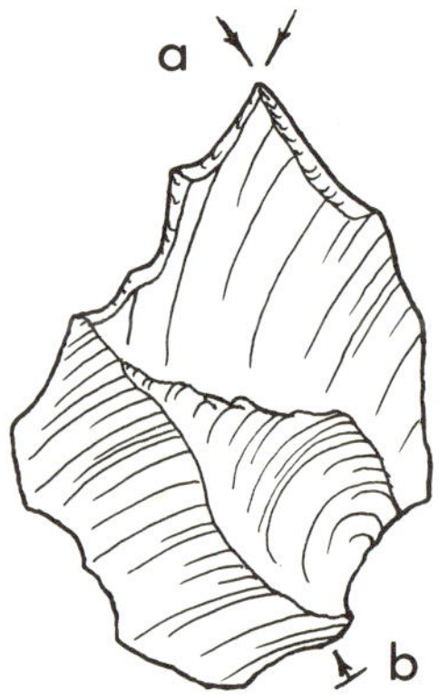

Figure 6: *a*: burin arrows; *b*: barred arrow.

Since an arrow is meant to show direction of burin blow, burins with three or four narrow parallel scars can cause a space problem for placing the necessary number of arrows. In such a case, draw the shafts of the arrows, angled properly in the correct locations, but set every other one back a little or forward closer to the piece (fig. 34, *d*), so that their staggered arrangement will allow room for their barbs, even though the shafts are by necessity close together.

Purely as a detail of style, arrows should be neat, with long elegant points and their barbs laid back close to the shaft--not tweaked outward at an angle or blackened.

Barred Arrow

Shorter than the arrows used with burins, a barred arrow (fig. 6) has a line (bar) across the back end of the shaft. It is the symbol used with the dorsal view of any tool to indicate the position of the bulb of percussion and the axis of detachment when those features are not, respectively, at the bottom of the drawing and approximately perpendicular to the base line.

The most common use of the barred arrow is with tools whose type dictates the orientation of the piece regardless of the original location of the bulb of percussion. An example is a perforator that was made on the proximal (bulb) end of a blade or bladelet (fig. 38, *m*). Since pointed tools are usually oriented with the point up, such a perforator would require a barred arrow above the drawing of its dorsal surface.

If a tool needs a barred arrow above it and also must have a distal view (for example, an endscraper made on the proximal end of a flake (fig. 14, *f*), draw the barred arrow above the dorsal view, then draw the distal view above the barred arrow.

The same idea of placement is followed in the case of a side-struck piece with a profile view: the barred arrow is placed nearest the dorsal view, between it and the additional view of the piece.

In the case of burins oriented for drawing with the bulb end up (for example, a dihedral burin made on the proximal end of a flake), place the burin arrows nearest the piece and the smaller barred arrow a little farther away (fig. 32, *h*).

An occasional tool will show that it was struck off-axis when detached from the core, making its general appearance decidedly off-balance and leaning far to the side if it were oriented for drawing in the standard way, with the bulb of percussion at the bottom of the picture. It is better to straighten the piece to a more balanced pose. Then place a barred arrow just outside the drawing to give the correct information as to bulb and axis of detachment (fig. 24, *h*, fig. 26, *f*).

There is a strong technological need for a barred arrow with a side-struck flake. The dorsal pattern demands a certain orientation, but when the piece is set in that normal-looking position, the platform and bulb of percussion are on a lateral boundary, making the axis of detachment run parallel to the base line. A barred arrow is essential with the dorsal view.

Note, however, that if a view of the ventral surface of any artifact is drawn, the dorsal view will have no need for a barred arrow; the information on platform, bulb, and axis is supplied by the ventral view itself (fig. 34, *d*, *e*).

Link Mark or Hyphen

Rule a single dash, like a hyphen, between the multiple views of a piece if the additional views sit to the right or left of the principal view (fig. 32, *a*, *b*, *g*). The purpose of the hyphen is to suggest gently to the eye that the views thus linked at the same height above the base line are indeed of the same piece. Do not use a link mark with views that are above, below, or oblique; their proximity to a principal view is sufficient.

All this is more a matter of style than universal rule. It has been made a part of this method in an effort to standardize a detail that affects the look of a page. Widely used, but distracting and somewhat objectionable, are dashed lines that fasten the top and bottom of one view to the next, or a distal or proximal view to a primary view. The repetition of heavy dashed lines marching this way and that catches the eye and contrasts too strongly with the delicate precision of the artifact drawings.

When all the views of a piece have been penciled, rule the link marks at a convenient gap determined by the shape of the artifact. Tailor their lengths to suit the size of the piece. A small bladelet and its profile might take a one-sixteenth-inch hyphen; a large core or hammerstone would look better with hyphens perhaps one-half inch long.

Avoid placing these link marks in the narrowest gap between views. Visual linkage is stronger if they are ruled in a wider space, and well above the midpoint between base line and top line. Attention to such details as this is cumulative in effect and contributes style and finish to a layout of drawings.

Tentatively, pencil the hyphens in place but postpone inking them until the drawings have been mounted in layouts for publication. Then ink all the hyphens on each layout at one time with a type of pen that can be used with a ruler. This timing ensures that a page full of drawings will have uniform link marks and that any minor adjustments can be made in their placement to keep the marks from lining up at the same level all across a row of different artifacts.

In the case of a short "partial view" placed beside a full-length primary view to show, for example, a bit of inverse retouch (fig. 40, *c*, *e*), use a small hyphen at a level suited to the small view. If there is also a profile view, place a hyphen between it and the primary view at a height suitable for the two full-length views.

Ticks

Ticks are single short dashes placed at opposite edges of a piece to indicate the points through which a cross section was made. Ticks should be ruled very near, but not touching, the primary view. Make them parallel to the base line if the section is transverse, perpendicular to the base line if the section is lateral, and shorter than the hyphens used between views (fig. 38, *f*, *k*).

The final part of this discussion of marks and symbols is concerned with treatments to use for specific features.

Use Retouch

The scars of use retouch occur singly or in small irregular groups along an edge and have the same color and texture (patination) as the rest of the flaked surfaces on the piece. They are damage caused when the tool was put to use. If the scars look fresh, they are not use retouch but recent damage, perhaps resulting from excavation or transport.

It is best to omit minor or scattered use retouch; there is a danger that it will be mistaken for purposeful retouch in the drawing. However, the importance attached to use retouch by the archeologist in charge determines the stringency of exclusion.

Extensive and pronounced use retouch on a cutting edge should be included in the illustration, but careful drawing must duplicate its irregularities and its characteristic scar shapes to avoid the appearance of true retouch (fig. 22, *c*).

Another type of occurrence that might be called "use retouch" and should be included in the drawing is the scar of an impact spall, encountered on some projectile points. It is usually a long narrow scar that originated at the tip of a point and interrupts all the flake scars in its path. If the characteristics of overlying position and rippled surface are drawn with the clarity that results from awareness of the feature's identity, it will be intelligible.

Snap Surfaces

When the surface of a snap is visible in the drawing, give it very slightly curved direction lines, placed to run the long way of the area (fig. 33, *b*, *g*). Dashed lines are not used with a snap.

An entirely different kind of "snap" surface is the twisted-off scar encountered on microburins (fig. 37) (which, incidentally, are in no way related to burins) and pieces such as geometrics made by the microburin technique. Basically, a microburin scar is merely the result of a step in producing a tool by a certain technique that involves shortening or pointing a piece of stone at a slant, on either end or both ends, by simply twisting off a portion of it, usually in preparation for further shaping by retouch.

The "twisting off" action originates either at backing or at a small notch created just for the purpose of making the snap take place at that spot.

Although microburin scars start at an edge (either backed or notched), they tend to curl well into view on the dorsal or the ventral surface of the artifact. This method of separating a blade or bladelet into two parts results, of course, in the creation of two new surfaces of stone, both of which are microburin scars. One will display a surface that is a "flake scar," and the other will be a "bulbar surface." Since the artifact being illustrated may have either, examine the microburin scar closely and select the correct kind of direction lines to draw into the area.

The force that separated the stone was applied at the backing or at the notch, and in a sense that spot became the "point of impact." Therefore microburin scars will have either a miniature bulb of percussion or the beginning of a flake scar at that point. The direction lines have to begin there and should be filled into the tiny area in such a way that the surface looks somewhat twisted, signifying the action that produced it.

In the case of microburins made with the use of a notch, at least some minute remnant of the notch remains and is a helpful clue in identifying the origination point of the microburin scar.

Break Surfaces

If the surface of the break is visible in the drawing, leave it blank. Extend dashed lines from each end of the break to indicate that part of the artifact is missing (fig. 5, fig. 31, *c*; fig. 34, *f*).

Second choice is to draw a series of straight parallel lines completely across the break surface, its short way. However, the latter treatment is more useful for very deep and shattered-looking hinge fractures that need a dark shadow.

One situation where parallel lines across a break surface are preferable to leaving the surface blank is the presence of broken pieces in a reconstructed core. Dashed lines, expressing "break," cannot be used, and the small area of a break's surface, among the congregation of pieces, would appear as a mystifying hole if left blank.

Another type of "break" is an area of fresh "bag-broken" scars (fig. 5; fig. 26, *k*) that are surface in character and do not materially alter the shape of the artifact. If the patch of fresh scars has interrupted and removed some retouch, enclose the patch with a single line and leave the area blank. If the accidental flaking occurs on a single large flake scar surface, it is considered inconsequential and should be omitted entirely.

Cleavage Surfaces

Treat a cleavage surface with fine parallel dashes of mixed random lengths, all slanting at the same diagonal angle, over the whole cleavage plane. Space the dashes to produce the tone of shading needed.

If a very large cleavage is broken up into several planes, use a different slant for the dashes on each plane and vary their spacing to reproduce the changes in light and dark from one plane to another (fig. 7, *a*). If the same cleavage is visible in a second view of the piece, use matching slants on the area in the second view, so that it will correspond logically to the first, but make allowance for the changed perspective and lighting.

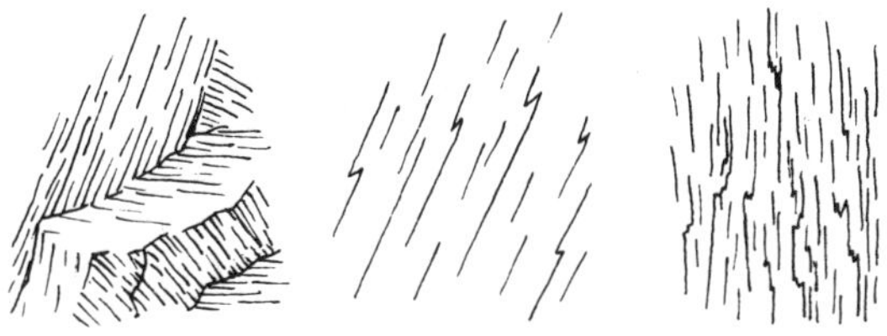

a b c

Figure 7. Cleavage surfaces. *a*: multiplane; *b*: quartz crystal; *c*: petrified wood.

A cleavage surface on that rare artifact of quartz crystal can be made to suggest its material by restrained and sparing use of a few zig-zag "lightning" marks mixed in with whatever scattering of random length dashes is necessary to tone the area (fig. 7, *b*). Maintain the same slant for all the marks.

A similar device can suggest petrified wood. In this case the direction of the dashes is determined by the grain of the "wood," usually lengthwise of the artifact. Instead of zigzags, mix into the dashes a few marks like stairsteps with very short treads and long risers, to give a splintery effect (fig. 7, *c*).

Fragments That Fit Together

An artifact broken during excavation or transport is glued together, and the joining is not indicated on the drawing. If the fragments were excavated as separate pieces, then later found to fit together, mark the joining with a heavy line across the drawing or use a heavy dashed line. If the pieces were found separately, they may well have different catalog numbers, which should be recorded to maintain their identities, even though they are drawn as one piece with the joining marked.

Details of the Raw Material

All such things as variegation of color, banding, seams of small crystals, and fossil inclusions should be ignored and omitted. Any attempt to indicate their presence muddles the information and clutters the drawing. The raw material from which a collection of tools was made is always discussed in the descriptive matter of the accompanying text for which the illustrations were drawn.

Thermal Fractures and Fire-Popped Pits

If heat cracks are a notable feature, draw the artifact in the normal way, then copy a few of the spidery branching lines where they run through some of the flake scars (fig. 8, *a*). Keep them very fine and do not use enough to compete with the direction lines.

Single smooth, round pits that look as if a hemispherical button had popped out of each one are a characteristic form of thermal effect (fig. 8, *b*; fig. 11, *d*; fig. 59, *b*). Another form that may be present on the same piece is a cluster of the pits, so crowded together that they look like a slice through honeycomb (fig. 8, *c*; fig. 44, *a*). Both of these forms of damage can be indicated by attaching small straight lines to the inside of each pit outline; aiming them all toward the deepest part of each one, use many little lines at the darkest side of its interior, interior, thinning them out to a very few shorter ones on the lightest side. The pits then look somewhat like craters.

An alternative treatment of the interior of the pits is a swirl of little lines in a whirlpool or vortex pattern (fig. 8, *d*).

All of this evidence of thermal damage should be drawn with restraint to avoid an emphasis that would make it dominate the drawing.

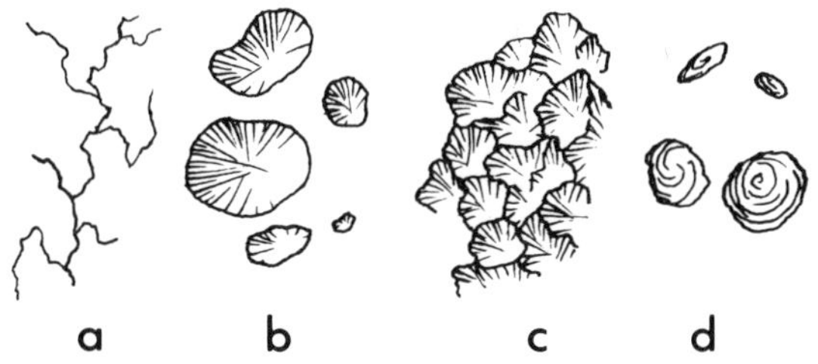

Figure 8. Thermal damage. *a*: heat cracks; *b*: single fire-popped cluster pits; *c*: cluster pits; *d*: swirl treatment of pits.

Furthermore, stay with one type of treatment for the pits; do not mix them on the same piece.

2 Customary Views and the Frameworks for Drawing Them

Artifacts are "posed" for illustration. The code of conventions dictates the orientation of tools and cores according to type and morphological details; but beyond that, specific poses for multiple views, in an orderly sequence of turns and alignments, are used to present artifacts' data with added clarity. The reader's perception of the illustrations gains firmer footing when the relationships of surfaces become reassuringly predictable and standardized by ritual repetition. Furthermore, there is increased confidence in comparisons of tools from different sites.

The set of common poses presented here is limited to basic ones that are most serviceable. Examples of them (fig. 9) will define the references to specific "views" throughout the text. In addition to their glossary function, the examples are meant to give beginning illustrators samples of possible views for drawing a particular artifact or tool type. Consult the many other illustrations for broader variety and demonstrated application. Use these views consistently, and they will contribute an overall appearance of order to a series of layouts.

Front and Back

There is no need to define such familiar designations as "front" and "back" except to mention their casual use in place of the technically more precise "dorsal" and "ventral" (bulbar surface) and to distinguish between views of the two opposed major faces of such artifacts as bifaces and cores. (Strangely, archeology's meanings for dorsal and ventral reverse biology's usage.) Some other poses, however, do need detailed explanations of their use.

Profiles

Although second in importance to a full-face image, a profile view of one's head, for instance, supplies many details that are not fully

Figure 9. Examples of the customary views of artifacts discussed in the text and shown in use throughout its collection of sample illustrations.

1: dorsal ("front"); 2: ventral ("back" or "bulbar surface"); 3; empty profile; 4: filled profile; 5: edge-only profile; 6: straight distal truncation view; 7: oblique distal truncation view; 8: distal view for an endscraper; 9: proximal truncation view; 10: butt view (here, Levallois platform preparation); 11: proximal view (here, a proximal transverse burin); 12: transverse (cross) section; 13: partial view; 14: tucked view (here, a partial view of the ventral surface of the tip).

The artifacts are a mixture of tool types. *a*: perforator; *b*: backed bladelet; *c*: double truncated blade; *d*: oblique distal truncation; *e*: notched and retouched blade; *f*: endscraper; *g*: multiple burin; *h*: retouched Levallois point; *i*: multinotch blade.

Sources: *h*: Marks, ed. 1976; *c*, *f*: Marks, ed. 1983; *a*, *b*, *d*, *e*, *g*, *i*: Wendorf and Schild 1980.

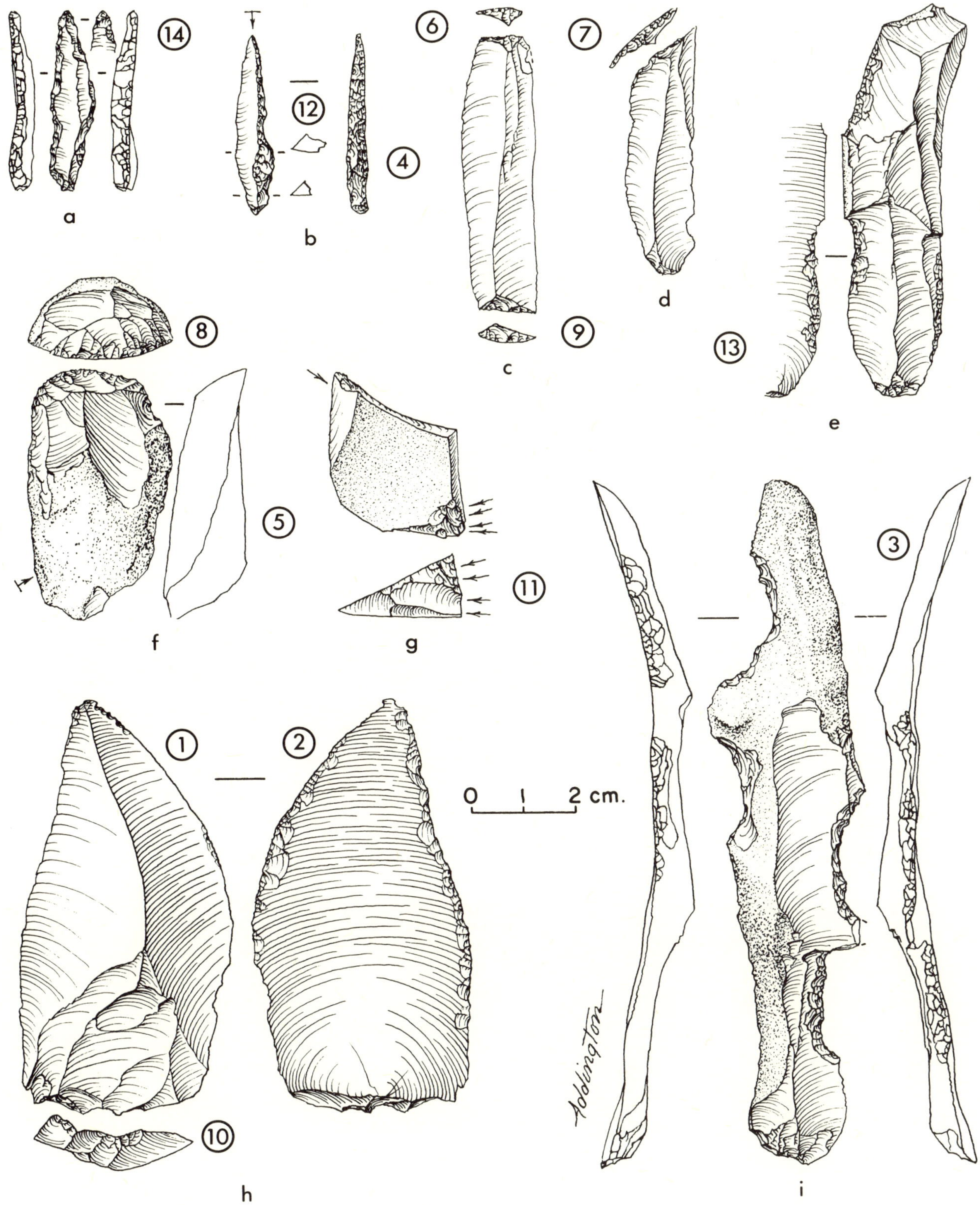

14
12
4
6
7
9
8
13
5
11
3
1
2
10
0 1 2 cm.
Addington
a
b
c
d
e
f
g
h
i

visible, or visible at all, in a front view--the slope of the forehead, projection of the nose, curve of the back of the head, and so on. The profile view of an artifact serves a similar purpose. It provides a better look at whatever flaking is too far around the side of the piece to be illustrated adequately in the drawing of the front, and it shows variations in thickness.

Pieces with backing or steep retouch on lateral edges need profiles to show all of the flake scars in those worked areas. An "empty profile", its scars outlined but not filled in with direction lines (fig. 9, *i*-3), is sufficient delineation when the force that made the scars was applied from only one direction. However, the profile view of a piece whose backing was produced by force from two directions must have the scars filled in with the appropriate direction lines to show that feature of its pattern of manufacture (fig. 9, *b*-4). Profiles of burins and cores should be as complete as their primary views.

An "edge-only profile" (fig. 9, *f*-5) is an outline of the piece from the side, plus a line that represents its lateral edge where any of the ventral surface is visible from that angle. All flake scars are left out of this simplified version of a side view. It can be drawn easily in very little time and incorporates a surprising amount of data. A piece with neither backing nor retouch that would require detailed drawing may still need its three-dimensional peculiarities clarified by an edge-only profile. This simplified view will show the artifact's variations in thickness down its entire length, usually a more informative drawing than a single cross section of the average artifact and drawn in considerably less time than a series of cross sections.

Draw a profile view so that it stands against a vertical guide line penciled for the purpose, touching the vertical only where the artifact touched the drawing paper when the piece was in position for outlining its front view. Few pieces are perfectly flat on the back; there is almost always some curvature or "bowing," often accentuated by a prominent bulb of percussion, that makes the ventral surface arch free of the paper (fig. 9, *a*, *b*, *f*, *i*). To the same degree, its profile will arch free of the vertical guide line. The character of the bulb, whether bulging or barely perceptible, and the angle of the platform at the butt are two more bits of data that the simple edge-only profile supplies.

Most endscrapers need a profile view to show the angle of the scraper and any undercutting or projecting hinge fractures of the retouch that forms the scraper end (fig. 12, *b*; fig. 16, *e*). A profile is especially desirable for a carinated endscraper (fig. 19, *f*; fig. 20, *i*) whose distal view can show the maximum thickness of its "keel" shape but not the decrease in thickness beyond the high point. Unless the archeologist in charge decides that time and layout space are too severely limited, supply at least the "edge-only" profiles as well as distal views (fig. 19, *d*), for endscrapers that do not need more detailed profiles, such as extensive lateral retouch would demand. That many of the endscrapers are displayed here without profiles simply shows by example that the ideal must give way to the reality of regrettable limitations.

Thin, flat endscrapers, often "ogival", whose retouch in not steep and can be seen well enough in the dorsal view (fig. 11, *b*, *c*) need only front and profile views.

Since backed pieces, in their many varieties, almost invariably require profiles, a member of that general category will be the most suitable to take through the following detailed steps of profile drawing.

To draw a profile view of a backed bladelet, for example, to accompany the front view already drawn, rule the appropriate selective horizontal guide lines (a procedure fully explained later in this chapter), including a base line and a top line. The first need is to gauge where to draw the profile with assurance of proper space between the views. Turn the paper so that the backed edge of the pencil drawing is nearest you. Hold the artifact, its backed edge fully visible and its ventral surface barely out of sight, alongside the backed edge of the primary view and about one-fourth inch away, (one-eighth inch if microlithic), fitting it in perpendicular position between the base line and the top line. Looking straight down on the bladelet held with its side displayed, make a tiny pencil mark at its back to tell you where to rule a vertical guide line that will allow room for the profile and the necessary space between the two views. Return the drawing to the normal position with the base line at the bottom and rule the vertical through the little mark.

With the vertical in place to act as a guide, the artifact's profile will be drawn to touch the vertical at the points where it touched the drawing paper while it was posed for its front view.

After ruling the vertical, again turn the paper so the backed edge of the bladelet drawing is nearest you. Fit the bladelet, in profile position, between the top and bottom guide lines so that it touches the vertical at the points described above. Holding the bladelet in position and looking straight down on it with one eye, draw the end that is not hidden by your fingers. Turn the paper around, hold the bladelet by its other end, and fit it back inside the part you have drawn and against the vertical. Then draw the rest of the profile outline.

The multiple horizontal guide lines make it a simple matter to transpose the scars of the backing from the primary view to corresponding positions in the profile. Five or six horizontals probably will be enough for a bladelet; a backed piece several inches long may need a dozen or more. A clear plastic ruler makes the horizontals no trouble at all. An artifact with scores of complex retouch or backing scars is no longer "too difficult" to draw in profile when enough horizontals, ruled to extend across both its primary view and the area of the paper to be occupied by the profile, cut the mass of scars into segments that are easier to copy without losing one's way and, at the same time, mark exactly where to draw the scars into the outline of the profile.

Truncations

Often, the flaking that performed the "cutting off" to create a truncation is so little visible in the dorsal view of the artifact that a direct look at the area of the truncation surface itself is necessary. Make this additional view show the outline of only the truncated area and its flake scars, omitting any other parts of the piece that may be seen from that angle.

If the primary view of the truncated artifact is the dorsal surface, place the additional view of the truncation itself so that the portion of its outline that represents the meeting of the truncated area with the dorsal surface of the piece is nearest the primary view (fig. 31). Note that this orientation is the reverse of that specified for distal views of endscrapers and is, therefore, part of the code that identifies the two different tool types.

If the general trend of the truncation, whether distal or proximal, is roughly parallel to the base line, extend vertical guide lines from each end of the truncated area of the primary view and draw the truncation view between the guides and about a quarter-inch from the main view--a little closer if the artifact is small. An exception to this spacing is a truncation that served as the striking platform for burin blows ("burin on a truncation"). In this case supply the burin arrows before drawing the truncation view, to be sure of room for the arrows (fig. 33, *e*, *l*, *n*).

If a truncation is so strongly oblique that to draw it between guides perpendicular to the base line would cause severe distortion, draw it in an oblique position, parallel to the truncated area of the primary view, between guide lines perpendicular to the general trend of the truncation (fig. 31, *j*, *k*, *l*).

Determine the general trend by a ruler laid across the drawing from end to end of the truncation; then, with a right-angle triangle set against the ruler, extend verticals from the ruler at each end of the truncation and draw the added view between those guides.

Earnest illustrators may have a tendency to "over-illustrate," in the sense that they will routinely draw a complete back view of every tool with any retouch at all on its ventral surface, such as a perforator with a few small scars at its tip. A partial view (fig. 38, *h*) of that detail, along with other complete views of the piece, is enough. An archeologist may be limited to one layout of illustrations to accompany an article. It would be wasteful of space to draw a complete back view of a large flake, for instance, when one retouched lateral edge is the only notable feature of the ventral surface. Partial views, in such cases, are adequate to display the information. Examples of partial views can be found throughout the manual.

One style of partial view that calls for some discussion is the tucked view (fig, 18, *a*; fig. 39, *m*, *n*, *o*; fig. 40, *b*, *e*), useful for severely limited layout space. Distal or proximal edge features, such as retouch and short burin facets on the back of a large piece, are shown by a drawing that appears to be "tucked" behind the primary view. Draw only the portion with the pertinent feature, plus a little of the neighboring ventral surface, in distal or proximal position as needed.

Use tucked views of lateral edges with caution. For example, a right (ventral surface) edge tucked behind a left (dorsal surface) edge can be confusing, if not incomprehensible. Furthermore, in a case of gross difference in outline--such as a deep notch tucked behind and coinciding with pronounced convexity--the space-saving aspect is lost and a layout problem created.

On the other hand, a lateral tucked view can make both the left and the right of a group of multiple views of an artifact equally strong and suitable for placing at either left or right boundary of the layout. The advantage can be a composition flexibility greater than that of an open-edged partial view, which ideally should neither begin nor end a row of drawings.

The rest of the examples of artifact views explain themselves. However, some points concerning the use of multiple views in special circumstances need to be aired.

A litter of oddly angled shapes surrounding a principal view causes confusion and must be puzzled out by the reader before the data can be understood. Numerals, dashed lines, or curved arrows to explain the various swivelings add clutter and are more appropriate with diagrams and schematics. If possible, keep to a single consistent system, with regard to angles of view and degrees of rotation, for drawing the "sides" of artifacts.

Use complex arrangements of views sparingly; indeed, save them for dealing with such legitimate problems as reconstructed cores or a group of burins that fit together (fig. 61; fig 34, *c*, *g*). When the purpose of the illustration is to show that a number of pieces were once a whole, spend some time planning not only which views are necessary, but in particular what placement of those views makes a logical, orderly arrangement that can be accommodated by the anticipated layout size and proportions. All the effort spent on drawing a spectacular "frozen explosion" of many parts is wasted if the "explosion" concept entails a chaotic proliferation of views and, equally undesirable, produces an indivisible arrangement of such proportions that total loss of detail and meaning would result from reduction to page size.

Illustrators' attitudes toward drawing multiple views of artifacts seem to range from "the least I can get away with," to "every surface lovingly delineated in its entirety." Something sensibly in between these two treats the material with more scholarly respect and handles the twin limitations of money and layout space with practical restraint.

One of the advantages of drawing artifacts instead of photographing them is the opportunity to pass along more information about a piece by using additional views to give the reader "a walk around it." Drawing an object from different angles seems to daunt some illustrators into providing one "front" view and letting the reader guess the rest. Space limitations are understandable. Nevertheless, it is a loss; a look at other surfaces could fill out the record. To encourage drawing as many views of an artifact as its individual characteristics call for, here are some helpful techniques and shortcuts--a system of guide lines that grew out of a need for the aid they furnish. If one factor had to be credited with saving time, increasing accuracy, and lessening aggravation in the illustrating of artifacts, it would be the routine use of a "selective grid" (fig. 10). With it, you can avoid many of the blatant flaws that suggest carelessness. For the front and back views of a piece to be grossly different in length and outline is inexcusable; yet, surprisingly, it is a common fault. Nor can one excuse flake scars that do not transpose, from one view to the next, at the same level. Accuracy in such elementary details as these is not dependent on artistic ability.

Early in this chapter, the discussion of how to draw profiles and truncation views described the use of some auxiliary guide lines; the major framework for drawing multiple views requires a more detailed exposition.

Selective Grid

A selective grid, as the term is used in this manual, is a network of horizontal and vertical guide lines, ruled lightly in pencil, their spacing tailored to the individual artifact, to aid in drawing successive multiple views of the same piece. The straight black lines in figure 10 are meant to represent what would be pencil lines in actual practice.

First, pencil the primary view of the artifact; then rule a horizontal base line and, parallel to that, a top line, touching the base and the top of the view. Extend the lines far enough to the left and right to provide a bottom and top guide for drawing whatever additional views are planned.

Next rule horizontal lines, parallel to the first two, right across the primary view and the area to be occupied by the additional views. Run these interior parallels through notable features of the artifact's outline, such as corners, points, and long curves, and also through some prominent surface features that will show up in the outline of the profile view, such as the highest point or a protruding hinge fracture.

Draw the profile view (instructions earlier in this chapter) to fit between the base line and the top line, with the aid of the other horizontals.

Next comes the back view. Holding the ruler against the base line, set the right-angle triangle to slide along it, and pencil vertical lines to form the selective grid. Draw the verticals to touch the left and right limits of the primary view, and to run through the features of the outline of the piece as was done with the horizontals. It is not necessary to run through the same features. The points at which all these grid lines touch or cross the outline of the artifact in its primary view can now be located accurately and will speed the drawing of the back view of the piece when you follow the rest of these instructions.

Since the outline of a back view is the reverse of the front view, the random spacing of the verticals of the grid must be reversed. It can be done quickly and easily without measuring. Lay the straight edge of a piece of paper *below* the base line of the front view and make a small mark on the edge where each vertical touches the base line. Then turn the scrap of paper around and lay it *above* the base line at the location where the back view is to be drawn. The ticks on the edge of the paper can be copied as dots on the base line, and their spacing will be the reverse of that for the front view's grid. With ruler and triangle, draw the verticals from these dots.

Figure 10. The selective grid. In practice, the grid lines are ruled only in pencil. After the pencil drawing of an artifact has been inked, grid lines are erased. A-A: base line; B-B: top line; C's: horizontals; D's: verticals.

In the upper diagram (a tool), X: first-drawn dorsal view; Y: second-drawn profile view; Z: last-drawn ventral view. If these same views of an artifact had to fit exactly a known layout width, the first-drawn would be X, then Z; then Y would be drawn between them.

In the lower diagram (a core), X: first-drawn front view; Y: second-drawn left profile; Z: last-drawn right profile. Note how the horizontals aid in transposing flake scars and features from one view to the next.

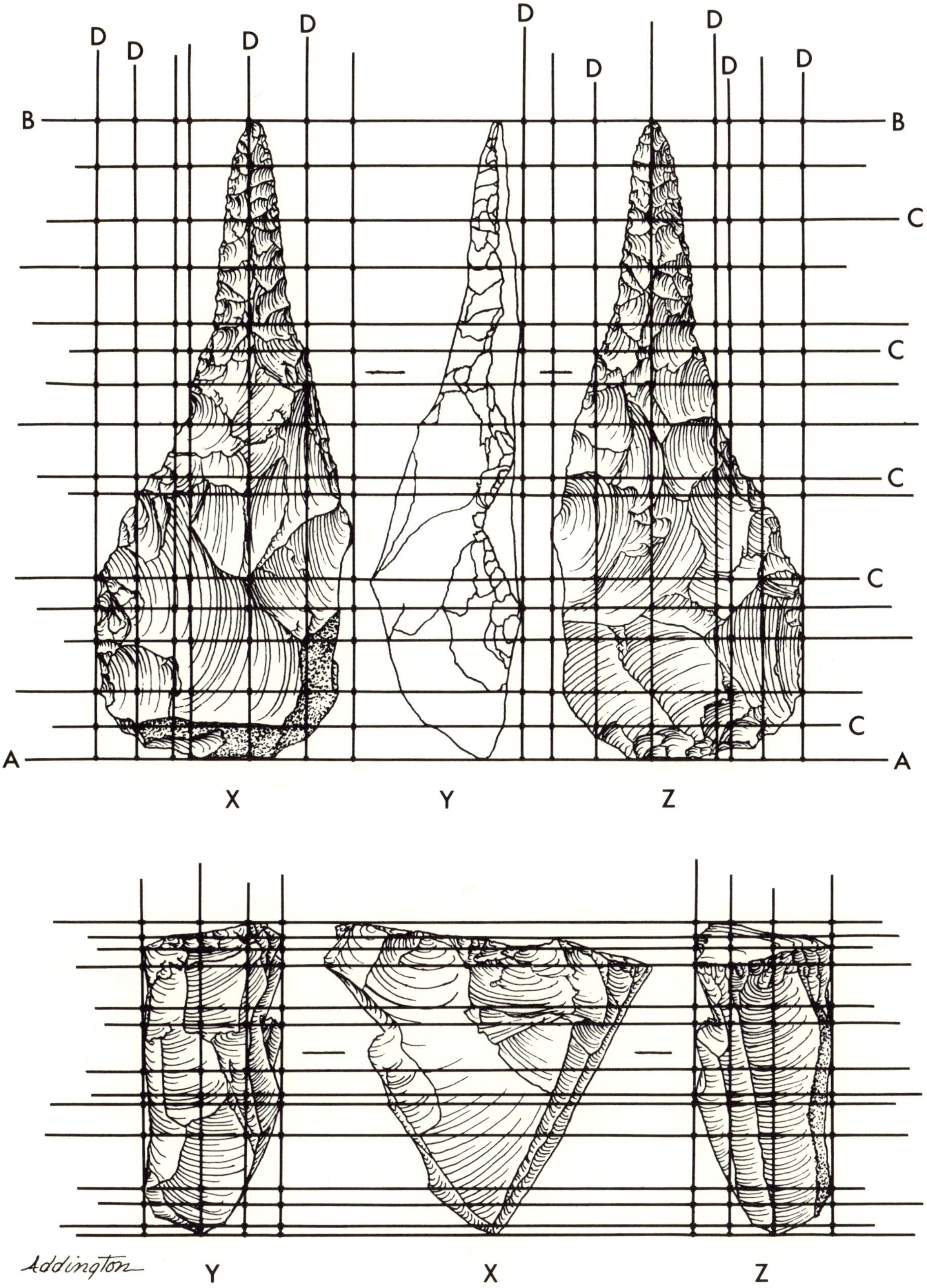
D
D
D
D
D
D
D
D
D
B
B
C
C
C
C
C
A
A
X
Y
Z
Addington
Y
X
Z

Now the points, corners, and such, of the front view can be located in reverse by means of the grid and the entire outline of the back view drawn quickly as an accurate reverse image.

Use the pencil with a light touch; the grid lines will have to be erased after the drawings are inked. The number of grid lines necessary is determined by the size, shape, and complexity of each piece. The horizontals are particularly valuable aids in transposing flake scars from one view to the next view where the same scars are visible, altered in shape by perspective, but not in height above the base line.

3 Orientation of Artifact Types for Illustration

To make reference use of this chapter convenient for the illustrator, the first part, dealing with tools, has been stripped of verbiage like a telegram and alphabetized. The information is general; understanding it depends on familiarity with the more thorough explanations elsewhere in this manual. Cores, the subject of the second part, need a less cryptic presentation.

Orientation of Tools

Backed Pieces

Orientation:	Usually bulb end at the bottom. Exception: proximal pointed (*see* ***Pointed Tools***).
Views:	Dorsal; profile of backed edge; ventral if inverse retouch present. "Natural backed" pieces may need cross sections.
Symbols:	None. Exception: proximal pointed need barred arrow.
Remarks:	Draw profile view of backing beside backed edge of the primary view--never beside opposite edge. Leave direction lines out of profile's flake scars if backing is normal (obverse); draw them in if the piece was backed from both directions (bipolar, bidirectional) or has inverse backing.

Bifaces (Handaxes)

Orientation:	Pointed (or narrow) end at top near center, as close to symmetrical balance as individual shape allows; propped for primary view so that in profile the general trend of lateral edge of piece will be perpendicular to base line, with the irregular line of lateral edge beginning and ending approximately on a single vertical.

Views:	Front; back; one or both profiles. Transverse sections through the tip if the distal portion is beveled or otherwise notable (drawn beside the tip at the level taken); and/or transverse section through possible backing or at widest part (drawn below, and aligned with, sectioned primary view).
Symbols:	None.
Remarks:	Biface profiles are commonly simple outline of the piece in that position, plus a line duplicating irregularities or zigzags of the "cutting edge" and the shape of backed area if present. A row of dots can continue from end of worked edge line, to mark the boundary between front and back "faces" in profile, when cortex encases lower part of piece and "cutting edge" is not full length.

Burins

Orientation:	General rule: burin blow directed downward. Exceptions: transverse burin keeps its transverse position; dihedral burin has point upward and arrows aslant; multiple burin keeps bulb of the piece at the bottom or is set so major burin's blow is directed downward.
Views:	Usually dorsal and profile or ventral and profile; additional views as needed. Burin on truncation often requires truncation view; transverse burin may need extra view of end involved.
Symbols:	Small arrow placed outside the outline of the piece, to show direction of burin blow.
Remarks:	Each burin blow that removed a spall and affects the present shape of the tip gets a small, neat arrow. Tentative short chipping scars do not rate arrows, nor do distal fragments of previous burin scars whose points of origination are gone.

Burin Spalls

Orientation:	Bulbar end at the top. Exception: a spall retouched into a perforator has point up; needs barred arrow if point is on bulbar end.
Views:	Dorsal; profile; ventral if retouch subsequent to detachment is present. Cross sections if needed.

Symbols: None.

Remarks: In deciding orientation, be aware that the bulbar surface formed by detachment of the spall during manufacture of a burin is the ventral surface of the spall. A remnant of the bulbar surface that belonged to the parent piece may be present on the spall, but its "direction" does not determine orientation of the spall.

Choppers

Orientation: Cutting edge across the top (or perpendicular to base line if size and shape are factors), so the working edge is well lighted. (Clinging to "use position," with cutting edge across bottom, merely obscures pertinent flaking by relegating it to the darkest shadowed area of the drawing.)

Views: Dorsal; profile or lateral section; ventral view if informative. Distal view if desired but without direction lines in flake scars.

Symbols: None.

Remarks: In distal view (or profile if worked edge vertical), a slightly heavier line for the cutting edge will clarify to which face of the piece the attendant retouch belongs.

Denticulates and Notches

Orientation: Simple: bulb end at bottom.

Views: Dorsal; other views as needed for steep or undercut retouch.

Symbols: None.

Remarks: Draw outline with care, giving full value to ins and outs, reproducing fresh sharpness or weathered rounding of projections as characteristic of the piece.

Geometrics (Lunates, Triangles, Trapezes)

Orientation: Bulb end at the bottom; cutting edge perpendicular to base line. Exception: scalene triangles place long point upward regardless of bulb position.

Views: Dorsal; profile of backed edge; ventral if inverse retouch present. Some broad, short triangles and trapezes need oblique views instead of profiles.

Symbols: None. Exception: scalene triangle with long point made on bulb end needs barred arrow above it (*see* ***Pointed Tools***).

Remarks: Omit direction lines in profile's flake scars unless backed from two directions (bidirectional) or inversely backed. If traces of sickle sheen are present, draw "repeat outline" of edge involved and indicate area of sheen with fine stippling, omitting all else within the outline (*see* "Stippling" in chapter 1).

Microburins

Orientation: Bulb end of what was the ventral surface of original whole piece at the bottom.

Views: Dorsal and ventral.

Symbols: None.

Remarks: Draw direction lines appropriate for the individual microburin surface (it can be either "flake scar" or "bulbar surface") so that the twist-off process is evident.

Pointed Tool *(worked to a point, such as* ***Perforators, Projectile Points, Scalene Triangles****)*

Orientation: Pointed end at the top.

Views: Dorsal; profile to show backing and/or thickness; ventral if there is inverse retouch.

Symbols: Barred arrow above if point is on bulb end.

Remarks: Profile of biface and bifacial projectile point: outline of profile, plus edge line of the piece (*see* ***Bifaces***).

Scaled Pieces

Orientation: Major opposed scaling at the top and bottom.

Views: Front; edge-only profile; back.

Symbols: None.

Remarks: Try to reproduce the characteristic scaly, shattered look of the scars.

Scrapers (Endscrapers and Sidescrapers)

Orientation: Single endscraper: scraper end at the top; double endscraper: major scraper at the top. Sidescraper: bulb end at the bottom and/or scraping edge approximately perpendicular to the base line.

Views: Endscraper: dorsal; distal; and/or profile with or without interior detail. Sidescraper: dorsal; profile or cross section as needed; ventral if inverse retouch present.

Symbols: Endscraper: position of distal view indicates "endscraper." Sidescraper: barred arrow if the piece was side-struck or struck off-axis.

Remarks: Endscraper: draw the distal view with the dorsal surface of the piece at the top. Sidescraper: draw the true shapes of the retouch scars; ink with care to avoid blacking out when reduced, especially if retouch is on shadowed right lateral edge.

Truncated Pieces

Orientation: Simple: bulb end at the bottom. Others: according to primary tool type. Example: a perforator made on the bulb end with the other end truncated is drawn with the point up.

Views: Distal truncation view is set above the primary view; proximal truncation view is set below. If primary view is dorsal, draw distal view with the dorsal surface of the piece on the bottom. If truncations are strongly oblique, draw distal and proximal views parallel to the truncations.

Symbols: Orientation of the truncation view (dorsal surface on the bottom) distinguishes truncation from endscraper.

Remarks: For the distal, proximal, or oblique view, draw only the area of truncation itself. In contrast, the distal view of an endscraper (and the proximal if a double endscraper) shows all of the piece that is visible from that angle.

When in Doubt

Orientation: Either the bulb end at the bottom or the "business end" at the top where it gets good lighting.

Views: Whatever views are needed to make the information clear.

Symbols: Follow the conventions that apply.

Remarks: Look up illustrations of similar artifacts in some of the suggested references.

Orientation of Cores

Most cores are oriented for illustration in a position that allows the direction of force that removed usable pieces of material to run from the top of the drawing toward the bottom (fig. 56; fig. 57; fig. 59; fig. 60; fig. 61). In the case of opposed platform cores, let the dominant (and therefore "top") platform be the one that produced the major pieces of usable material.

This rule of downward force applies to these core types:

-- Single platform
-- Opposed platform
-- Changed orientation or "Ninety-degree"
-- Pyramidal
-- Globular or multiple platform

The core types to which this rule does not apply are:

-- Discoidal (struck from all sides toward the center)
-- Levallois
-- Halfa

For discoidal cores, select as the up-and-down of the front view an orientation that will reveal platform angles in the profile view. Then draw front, profile(s), and back as needed.

Two kinds of specially prepared cores, Levallois and Halfa, are oriented, by convention, with the major flake scar's point of impact at the bottom of the drawing, not the top (fig. 58). Both types require front, profile, and back views.

An occasional core (usually single platform) can be illustrated adequately with simply a front view and a profile. Most cores need three views; some need four. Globular cores, including those called "multiple platform, patterned" and "multiple platform, unpatterned," usually need four views to show enough of their different surfaces and platforms.

Each adjacent view of a core represents a one-quarter rotation of the piece around its vertical axis and in sequence: for example, front, right profile, back; or left profile, front, right profile.

To draw most cores, level the face of the core to be as nearly parallel to the surface of the paper as possible, prop the piece with bits of eraser or pellets of dried rubber cement, and draw it in that position according to the instructions on penciling. Rule a base line, top line, and the rest of the selective grid as detailed earlier.

Bring the side of the core into position for a profile view. Now the original "face" of the core should be perpendicular to the base line on which the first view sits, and the various features through which the selective grid has been run must be matched with whatever corresponding features are visible when the piece is in profile position. Draw the profile, following the instructions for profile views.

Continuing the sequence for the back view, the position of the original face will be leveled, face down, on the paper; but do not attempt the nearly impossible job of drawing from the piece itself a perfect matching outline that is the reverse of the first drawing. Instead, follow the instructions about reversing the selective grid for a back view. When you have completed the reverse of the outline with the aid of the grid, then fit the piece into the outline, locate the interior features by the method discussed in the next chapter, and finish drawing the back view.

This system of positioning and rotation, applied consistently, makes the platform angles obvious and easy to compare with those of other illustrated cores.

Pyramidal cores are an exception to this "face leveling": the surface of the platform at the top of the core is placed so it is nearly parallel to the base line on which the point of the inverted pyramid rests in the drawing. It is probable that there will be enough irregularity of the platform to make a bit of its surface visible in one or the other of its profiles or in its back view (fig. 59, *a*).

An important reminder: A feature visible in more than one view must be drawn at the same level above the base line in those views, or the drawings will not be accurate portrayals of rotation around the imaginary vertical axis that runs through the center of the piece. The prescribed horizontal guide lines are more than helpful; they are essential if drawing is to proceed at a good pace and without a lot of erasures and corrections.

Refer to the discussion of core problems in chapter 6, on layout composition, to be aware of additional factors that should be considered before drawing a core.

Figure 11. Endscrapers. *a–c*: ogival endscrapers; *d*: massive endscraper showing old patinated flake scars, cortex, and thermal damage pits; *e*: double endscraper; *f*: endscraper on a retouched piece whose long major flake scar on the left demonstrates the placement and spacing of direction lines to make a scar twist, "climb," and ripple from proximal to distal end. Note the prescribed orientation of distal views of endscrapers: convention decrees that the "front" surface of an endscraper be at the top of its distal view. Contrast with distal truncations for the opposite convention.

Sources: *a–d*, *f*: Marks, ed. 1976; *e*: Marks, ed. 1977.

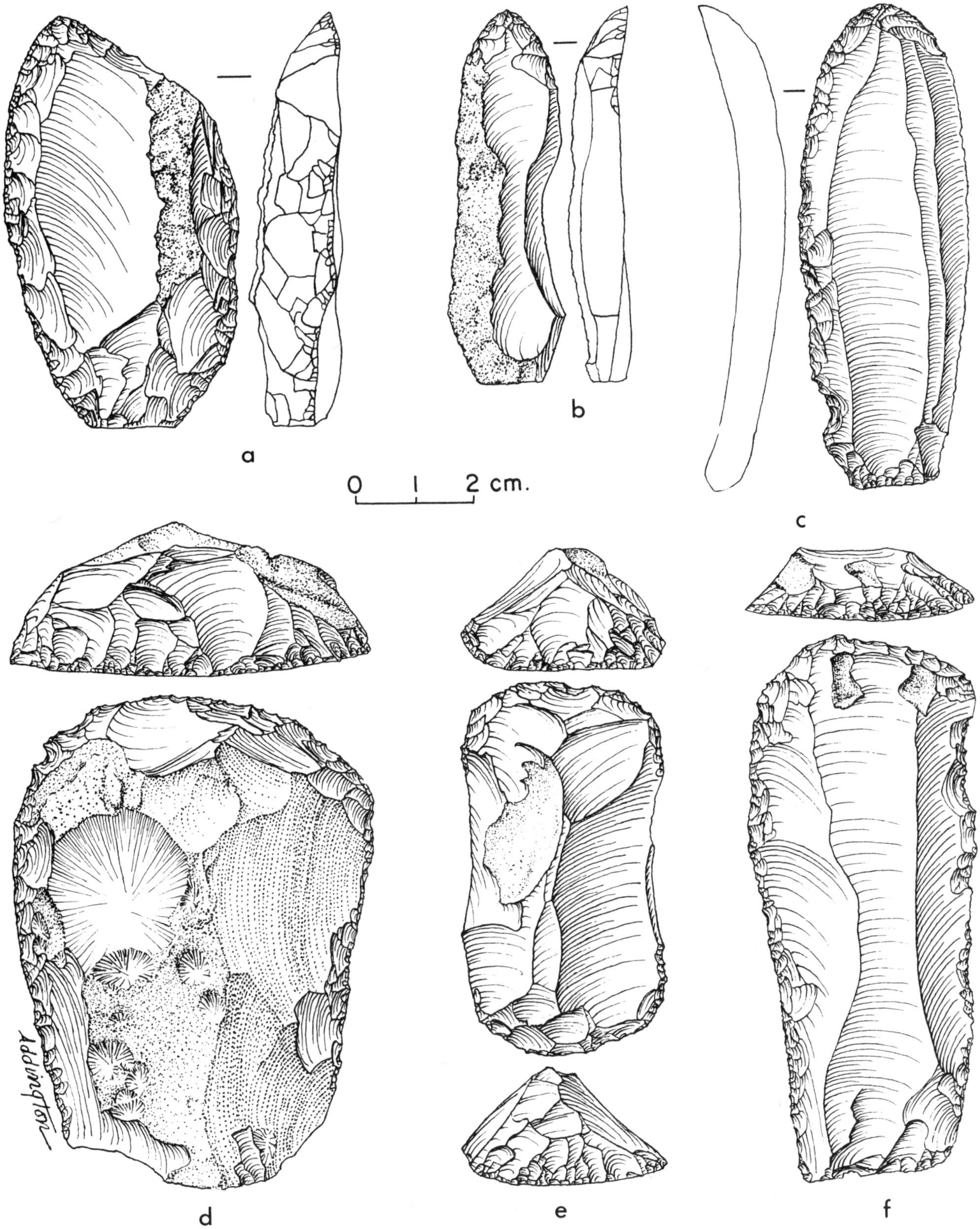
a
b
c
d
e
f
0 1 2 cm.
Addington

Figure 12. Endscrapers with some problem contours or unusual features. *a*: steep left lateral scar becomes a problem in the distal view; *b*: needed butt view of its Levallois platform preparation; *c*, *d*: endscraper/burins; *e*: bilateral endscraper with bifacially thinned butt for which a complete ventral view was used instead of a proximal tucked view to avoid the layout problem the latter can cause; *f*: edge-only profile is sufficient for this flat endscraper on a primary flake; *g*: shows treatment of large scooped flake scar on top surface of distal view; *h*: spacings of direction lines in top surface flake scars of distal view help make them appear to recede; *i*: endscraper, with a bit of inverse sheen on tucked partial view of ventral surface, has barred arrow to locate bulb of percussion and axis of detachment because its pose was straightened for drawing. Again, note orientation of distal views of endscrapers contrasted with those of truncations elsewhere.

Sources: *a*–*d*, *g*, *h*: Marks, ed. 1976; *e*, *f*, *i*: Marks, ed. 1983b.

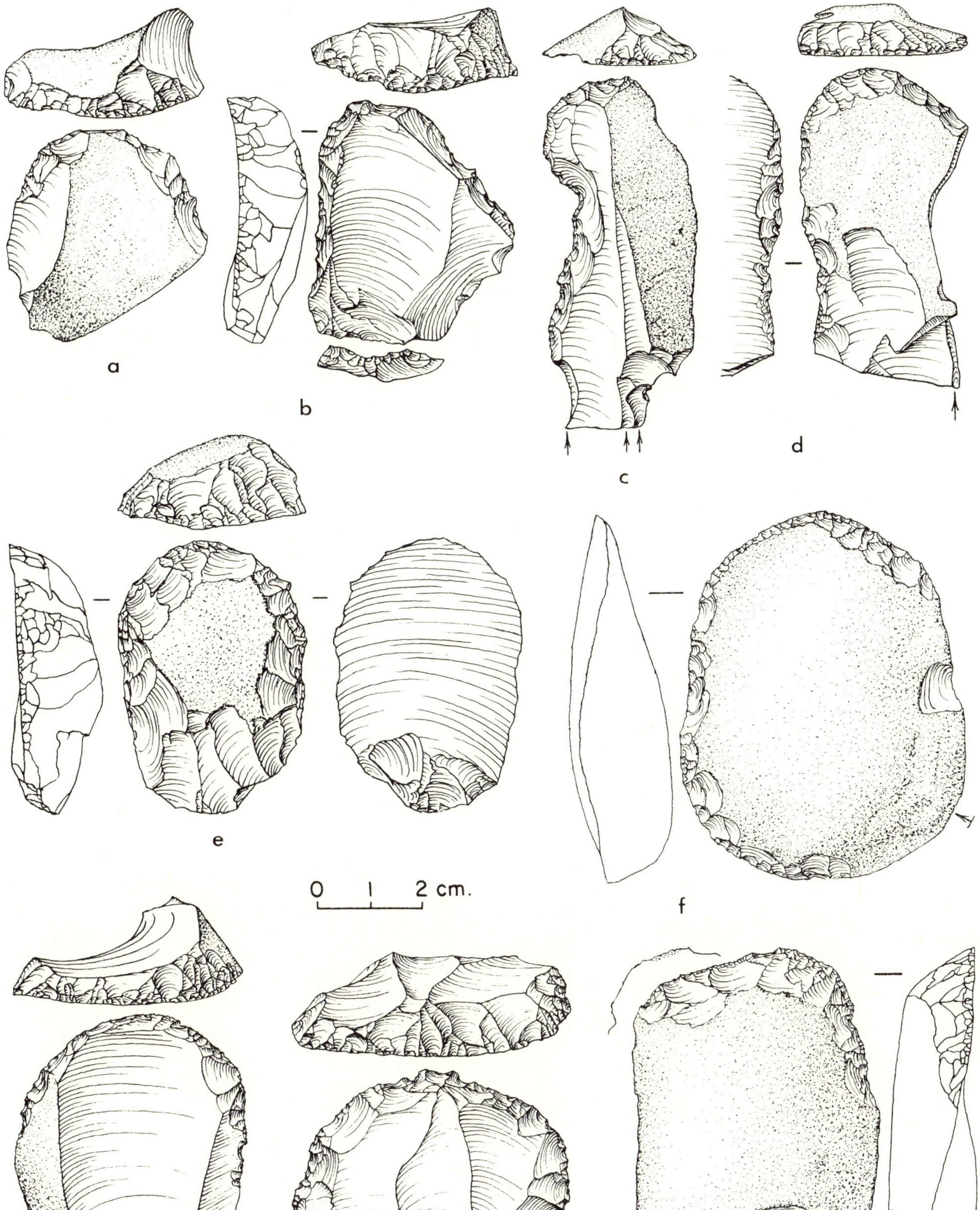
a
b
c
d
e
f
0 1 2 cm.
g
h
i
Addington

Figure 13. More varieties of endscrapers. *a*: endscraper on a flake; *b*: has uncommon cortex butt; *c*: endscraper on a blade with abrasion indicated by graduated dots; *d*: endscraper on a blade fragment; *e*: long right lateral scar drawn to show its origin was on the core and not on the piece; *f*: bilateral endscraper; *g*: lateral endscraper with opposed notch; *h*: upper central projecting peak in primary view gives nice problem for solution in distal view (note what perspective does to direction lines in big scooped scar when it is drawn in distal view behind the peak); *i*: bilateral endscraper with problem of severe perspective effect on big flake scar just visible in distal view.

Sources: *d*-*h*: Marks, ed. 1976; *a*, *b*: Marks, ed. 1977; *i*: Marks, ed. 1983b; *c*: Wendorf and Schild, assemblers, and Close, ed. 1984.

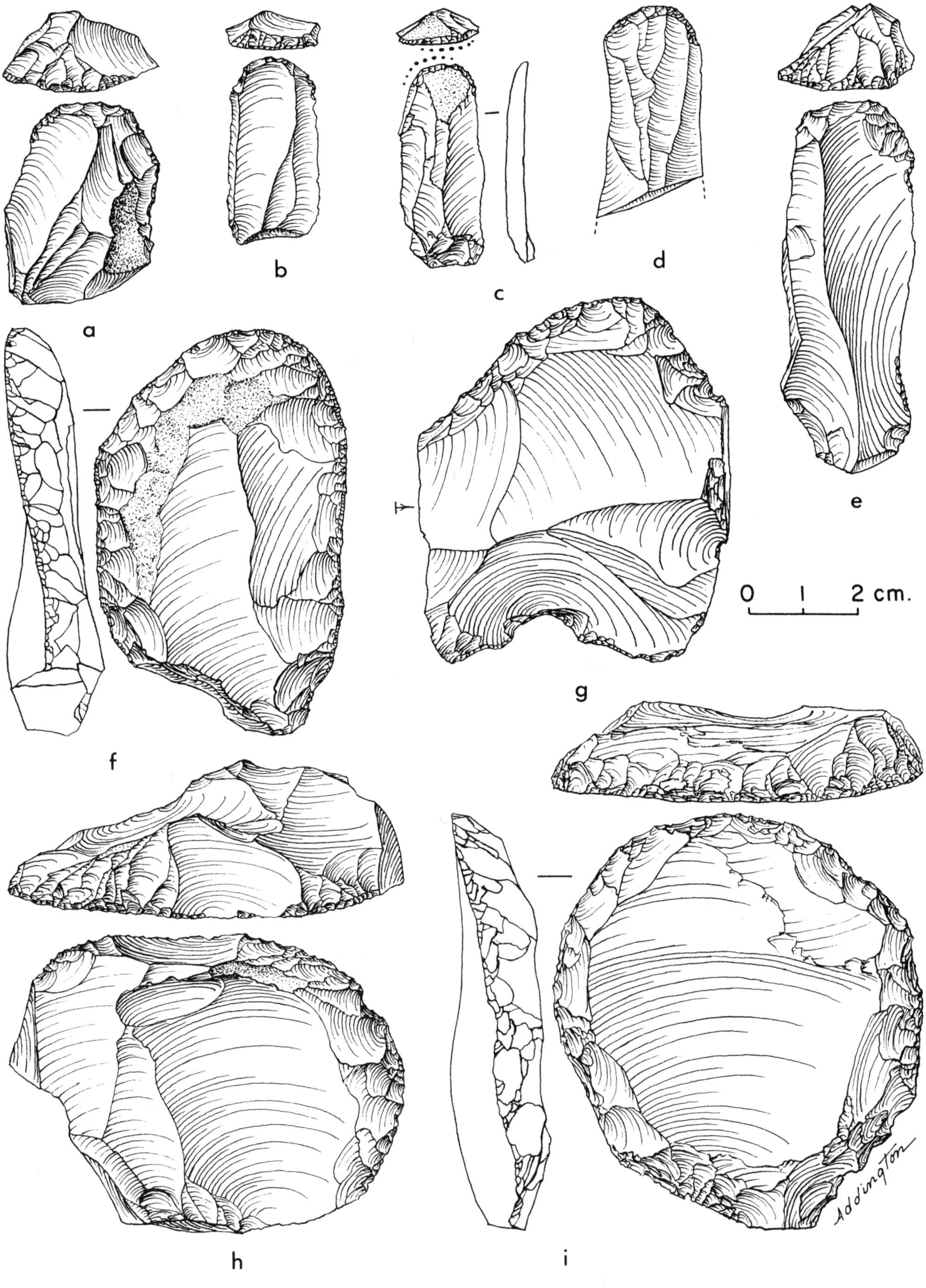
a
b
c
d
e
f
g
0 1 2 cm.
h
i
Addington

Figure 14. Additional examples of endscrapers. *a*: inverse endscraper; *b*: endscraper with uncommon bilateral pattern of flake scar remnants whose points of impact were removed by the long central scar before the scraper retouch was effected; *c*: endscraper on a primary (meaning mostly cortex-covered) flake; *d*: endscraper on a blade shows a few patinated scars; *e*: endscraper on a core trimming whose scars originated on a core; *f*: endscraper made on bulb end of the flake as barred arrow indicates; *g*: endscraper on a bilaterally retouched blade.

Sources: *a*, *b*, *f*, *g*: Marks, ed. 1976; *c*, *d*: Marks, ed. 1977; *e*: Marks, ed. 1983b.

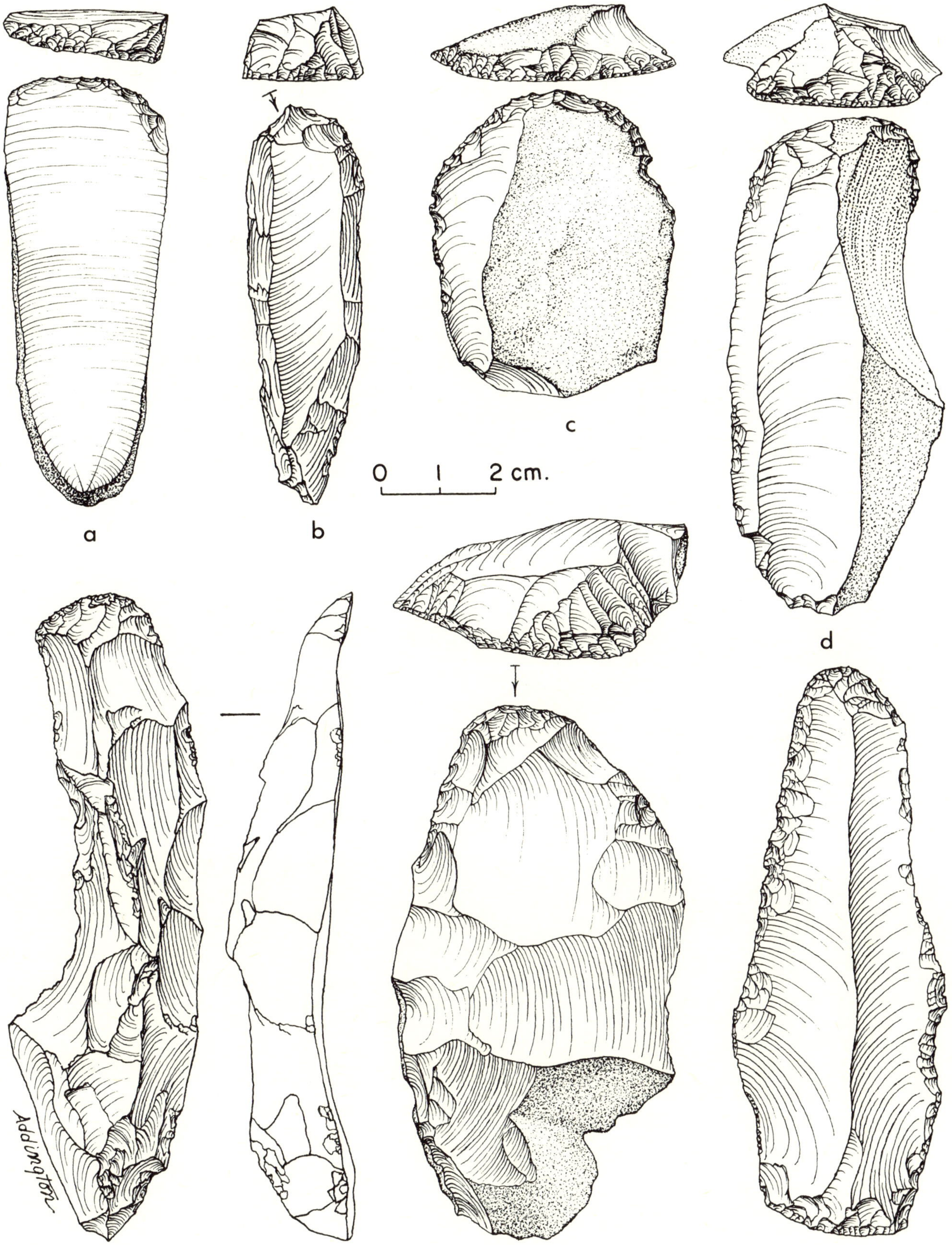
a
b
c
d
e
f
g
0 1 2 cm.
Addington

Figure 15. Denticulate endscrapers and more problem contours. *a–c*: denticulate endscrapers; *d*: endscraper on a retouched blade; *e*, *f*: Ksar Akil scrapers, a unique type whose delicate and precise denticulation requires careful drawing; *g*: unifacial celt-like piece; *h*: scraper on a side-blow flake with problem contour distal view that shows placement of direction lines to imitate the cupped and flared surface of the big scar, distorted by extreme angle of perspective.

Sources: *b*, *d*: Marks, ed. 1976; *a*, *c*: Marks, ed. 1977; *e*, *f*: Marks, ed. 1983b; *g*, *h*: Wendorf and Schild 1980.

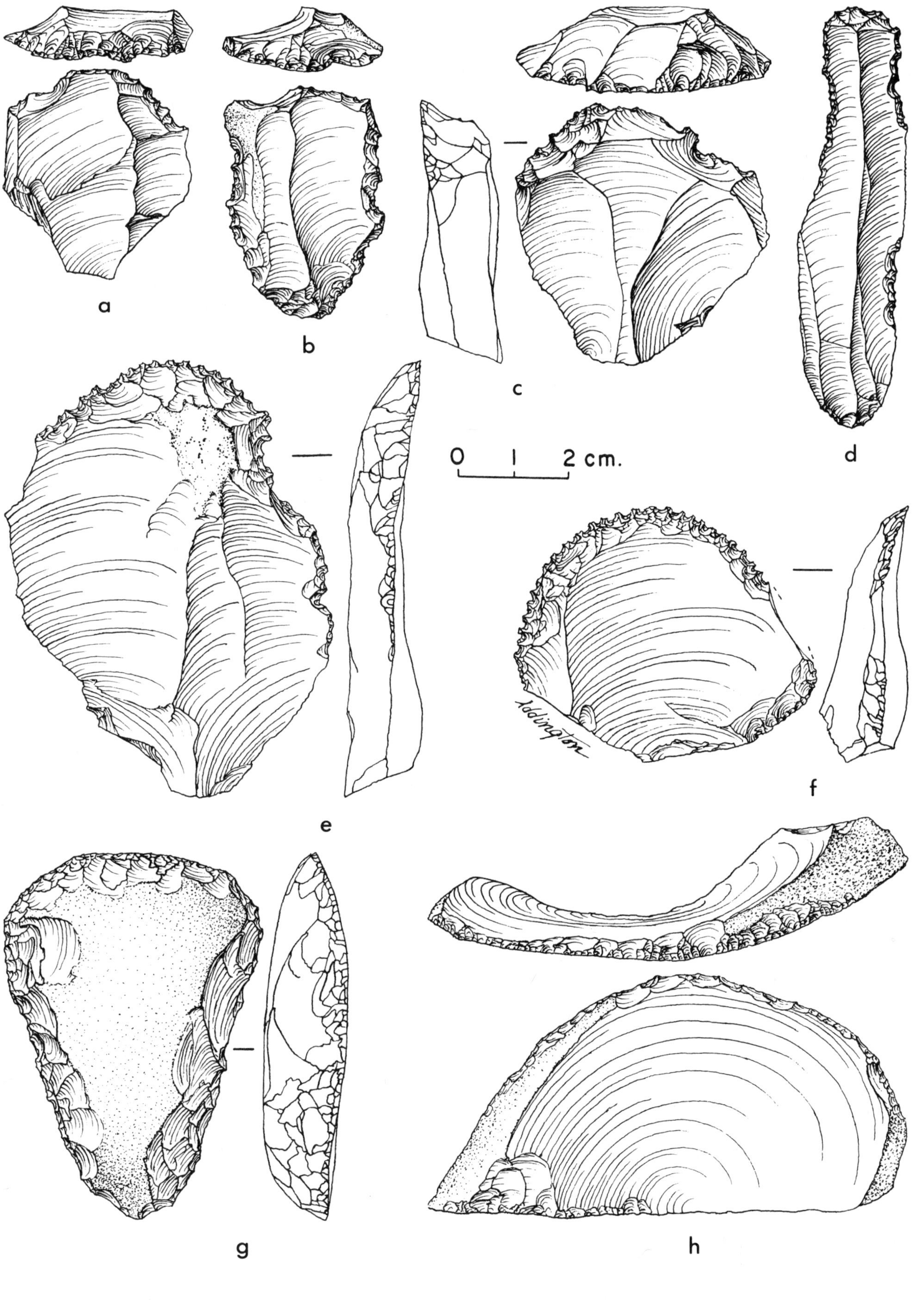
a
b
c
d
0 1 2 cm.
e
f
g
h

Figure 16. Thick endscrapers and nosed endscrapers. *a*: steep, thick endscraper with needed empty profile; *b*: endscraper on a retouched piece (made on proximal end); *c*: thick nosed endscraper with barred arrow showing axis of detachment from core; *d*: double thick nosed scraper whose distal and profile views handle the problem of a very steep-sided piece; *e*: thick endscraper; *f*: thick endscraper with some patinated flake scars; *g*: thick nosed scraper with clarifying profile and distal views; *h*: thick shouldered scraper with three-dimensional effect in stippling of cortex achieved by darkest shadowing alongside the lightest area and some darkening near left lateral edge to "push" it downward.

Sources: *a*, *b*, *d-h*: Marks, ed. 1976; *c*: Marks, ed. 1983b.

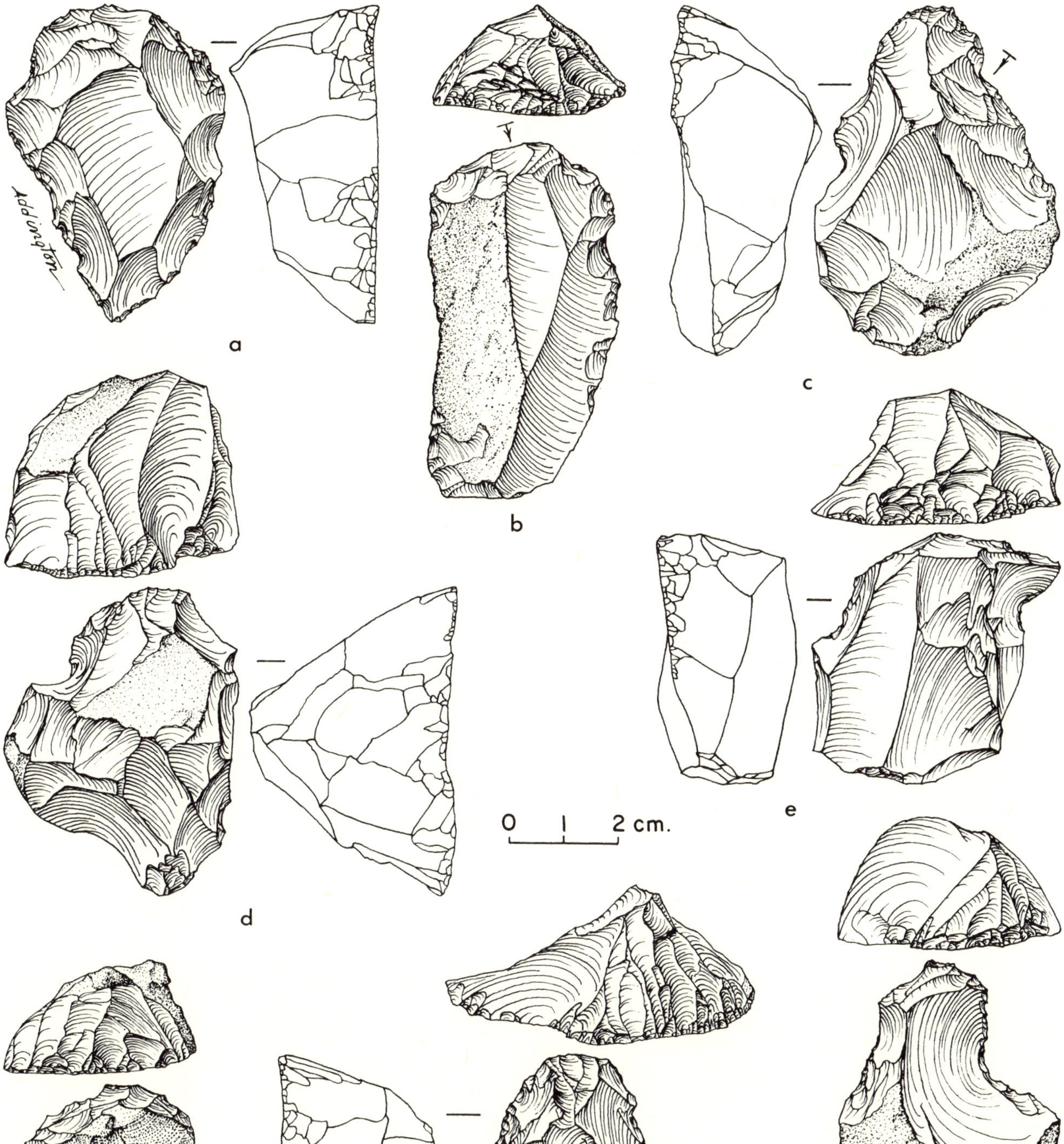
a
b
c
d
e
f
g
h
0 1 2 cm.
Addington

Figure 17. Small and thin endscrapers. *a–l*: various small or thin endscrapers; *m*: notched endscraper on a blade with inverse retouch; *n*: endscraper on a retouched piece with example of distal view that makes visible major scar recede by placement of direction lines; *o*: thin endscraper; *p*: unilateral endscraper whose distal view is example of lateral projection that is wider than scraper end of the piece.

Sources: *b*, *e*, *f*, *j–l*, *n*, *p*: Marks, ed. 1976; *a*, *c*, *d*, *i*: Marks, ed. 1977; *g*, *m*, *o*: Marks, ed. 1983b.

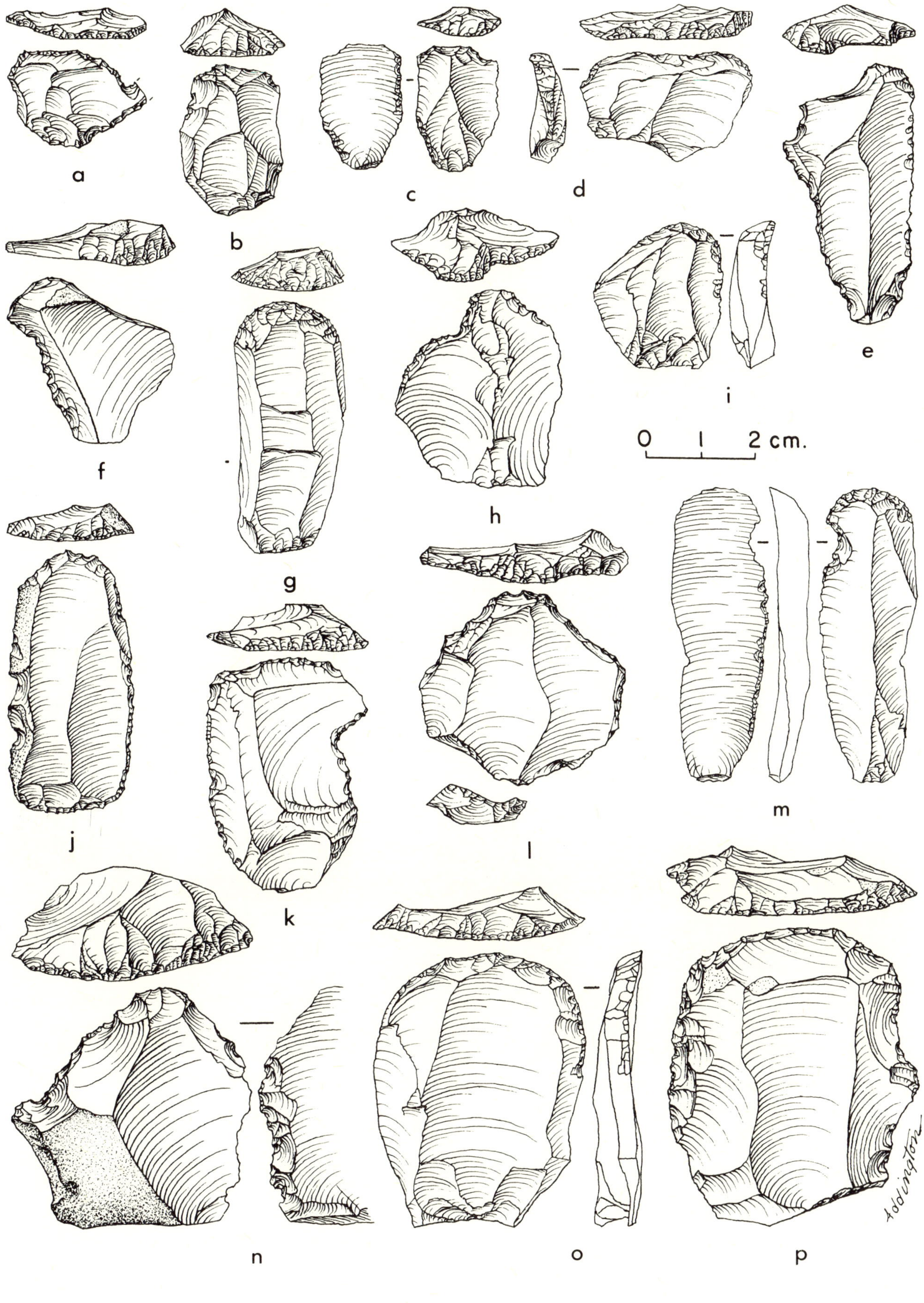
a
b
c
d
e
f
g
h
i
j
k
l
m
n
o
p
0 1 2 cm.

Figure 18. Endscrapers on reused Levallois flakes. *a*: ideal use of tucked view to show proximal thinning; asymmetrical bulge of ventral surface shown as blank area in distal view of *a* and *c*; on old patinated surfaces, rows of dots outline the scars, form the direction lines in them, and indicate the bulbar surface of *a*; dots in edge-only profile view mean old dorsal and ventral surfaces meet there, while solid line is used for edge formed by "new" retouch; old patinated flake scars are indicated in both major and butt views of *b* and *d*, with cortex in central portion of *b*. Note that position of butt view is in vertical alignment with location of the platform of *b* and *d*.

Sources: *a*–*d*: Wendorf and Schild, assemblers, and Close, ed. 1984.

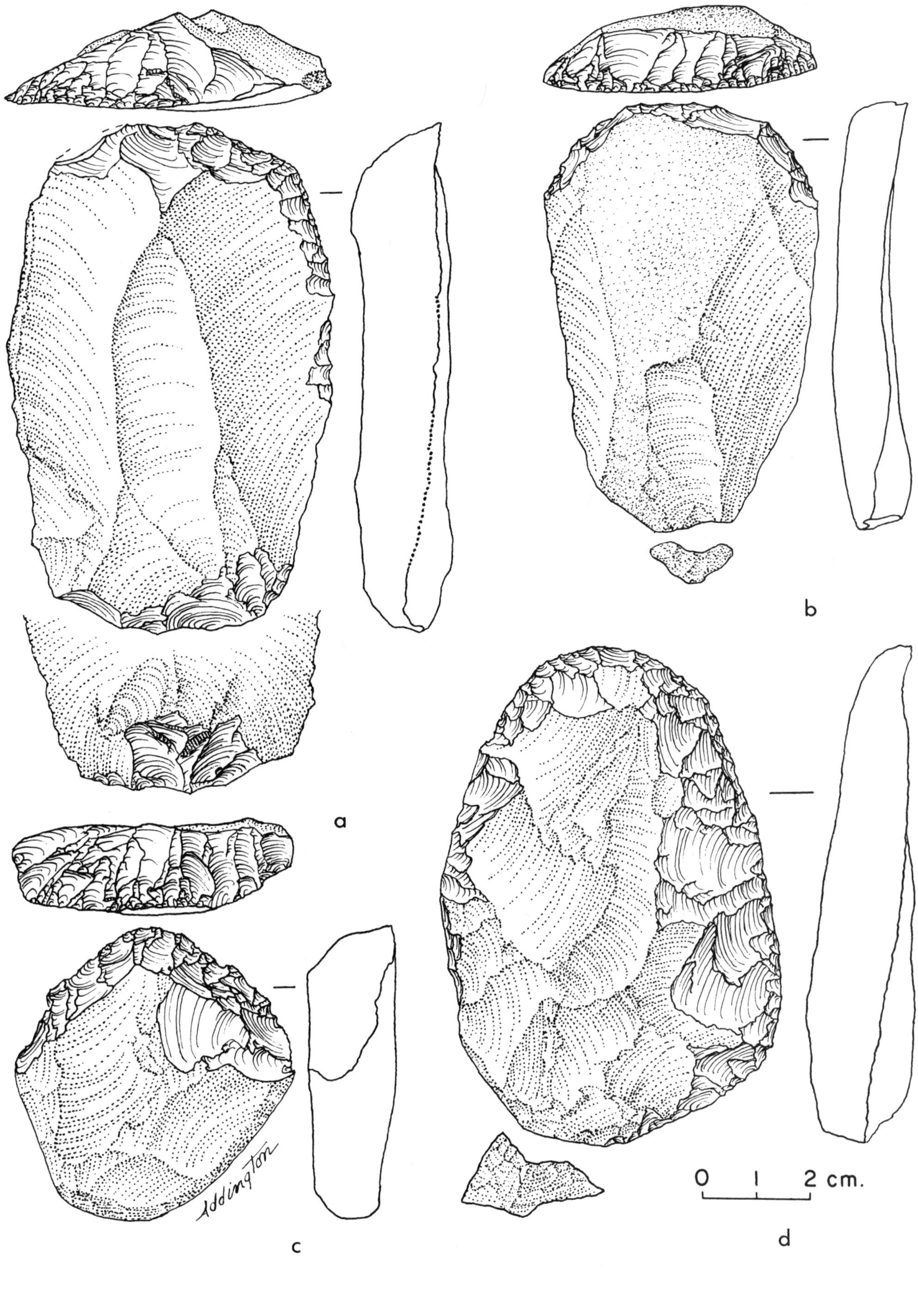
a
b
c
d
0 1 2 cm.
Addington

Figure 19. Carinated and semicarinated scrapers. *a*, *b*, *d-h*, *k*: carinated scrapers; *c*, *i*, *j*: semicarinated. Note, *a* would have benefited from an empty profile to explain that extreme convexity of the underside forced the upturned effect of the distal view, in this case left blank for clarity.

Sources: *a-g*, *i-k*: Marks, ed. 1976; *h*: Marks, ed. 1977.

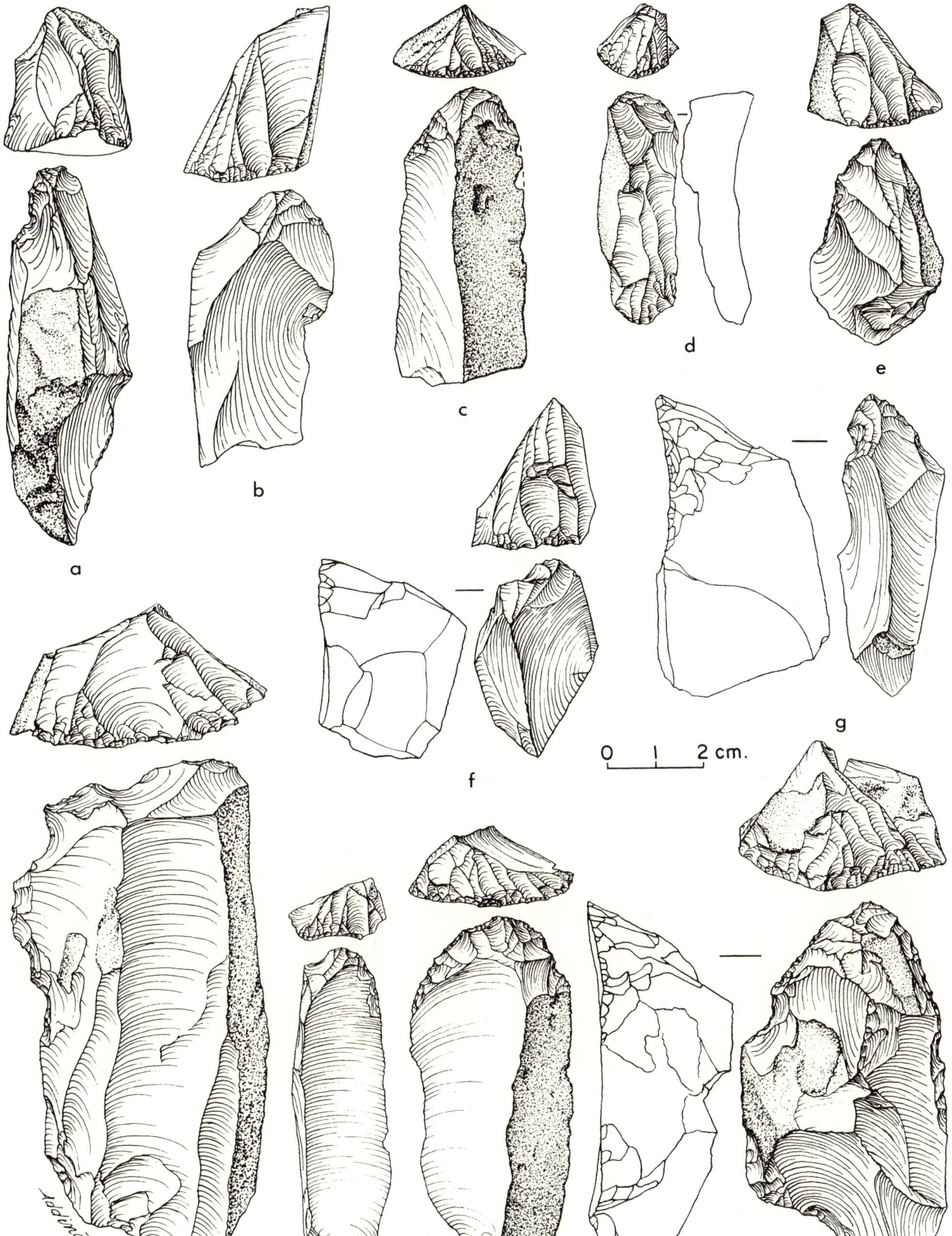
a
b
c
d
e
f
g
0 1 2 cm.
h
i
j
k

Figure 20. Lateral carinated scrapers and other variations. *a*, *f*: the bidirectional retouch that shaped them required filled profiles; *b* and *f* are inverse; *f* is made on proximal end of flake.

Sources: *a*-*g*, *i*: Marks, ed. 1976; *h*: Marks, ed. 1977.

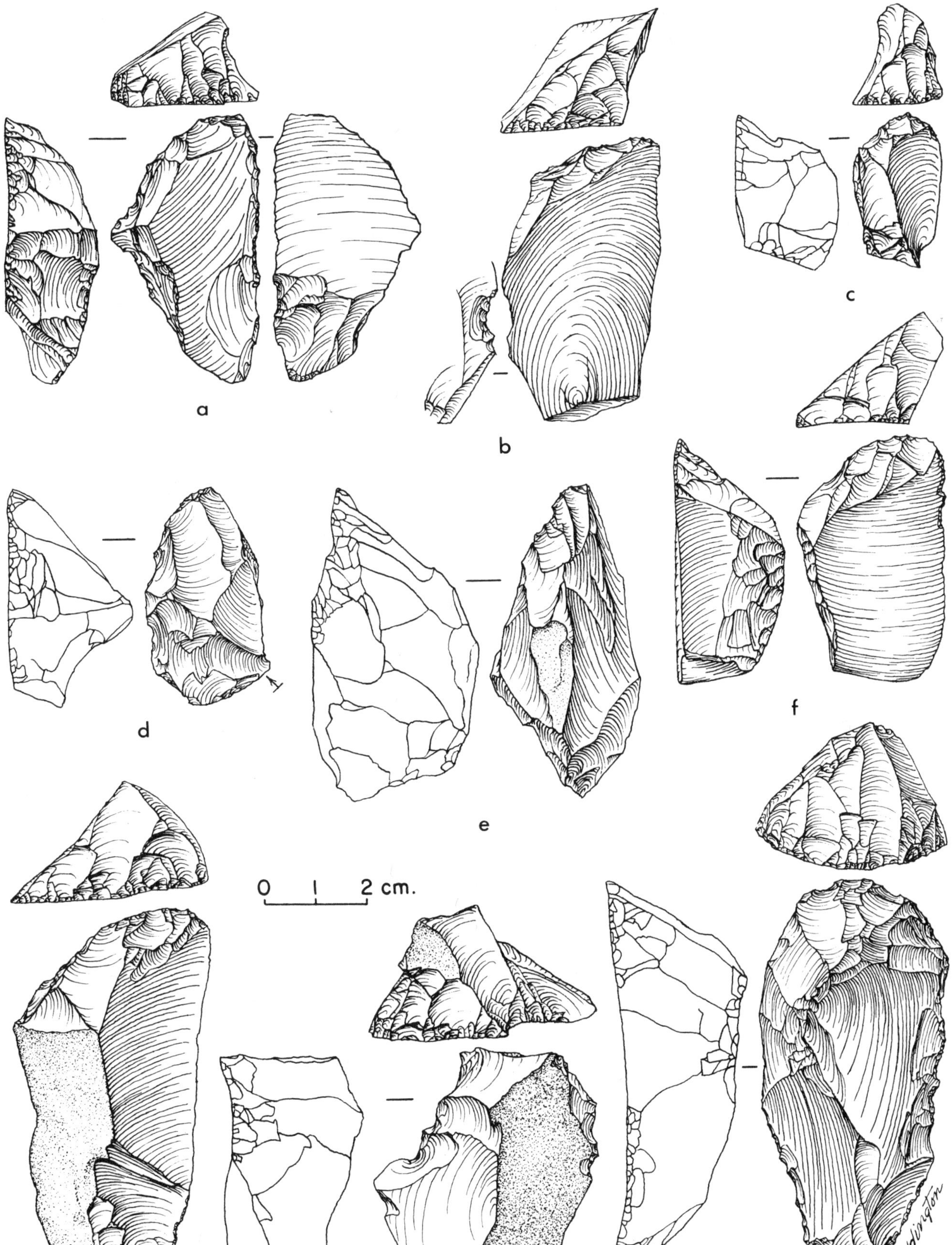

a
b
c
d
e
f
g
h
i
0 1 2 cm.
Addington

Figure 21. Examples of sidescrapers. *a*, *c*: transverse sidescraper and double transverse sidescraper. The prominence of the bulb on the ventral surfaces of these tilted the pieces so far toward the distal end that no distal retouch would have shown in the primary view without propping the distal end in a slightly lifted position for the major drawing; therefore the distal view of each exposes some of the ventral surface, here left blank to avoid interfering with the scraper retouch. *b*, *d*, *e*, *g*: unilateral sidescrapers; *f*, *h*: bilateral sidescrapers; *e*: denticulated sidescraper on an old flake; *h*: example of useful tucked view of distal features on the ventral surface.

Sources: *b*–*e*, *g*: Marks, ed. 1976; *a*, *h*: Marks, ed. 1983b; *f*: Wendorf and Schild 1980.

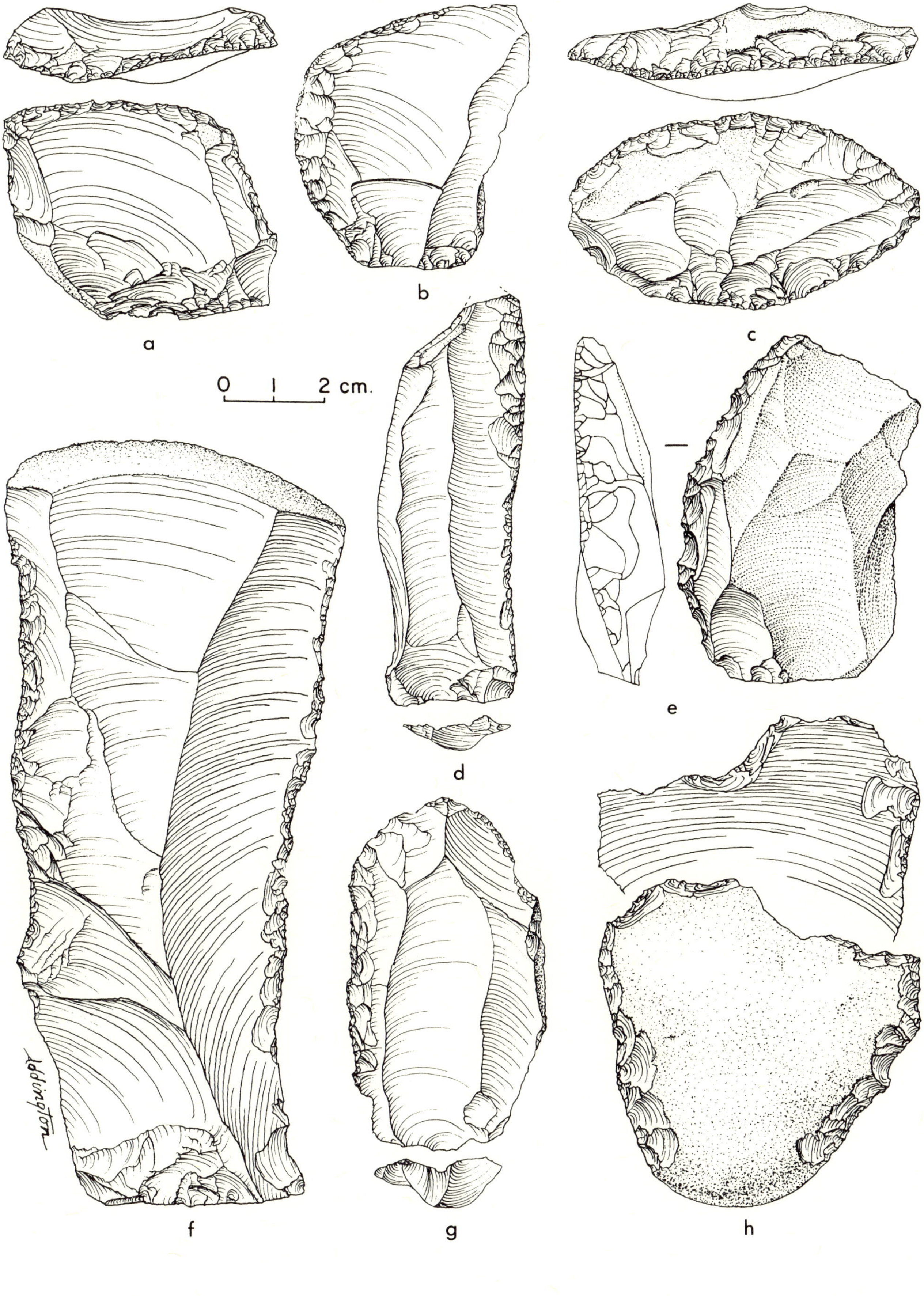
a
b
c
0 1 2 cm.
d
e
f
g
h

Figure 22. Sidescrapers and knife. *a*, *d*, *f*, *g*: double sidescrapers; *b*: convex sidescraper; *c*: natural backed knife; *a*, *b*, *e*–*g*: Levallois pieces with customary characteristic butt views that are in vertical alignment with the butt of the primary view and retain the tilt, dip, level, and such, of the bottom edge of the dorsal surface.

Source: *a*–*g*: Marks, ed. 1976.

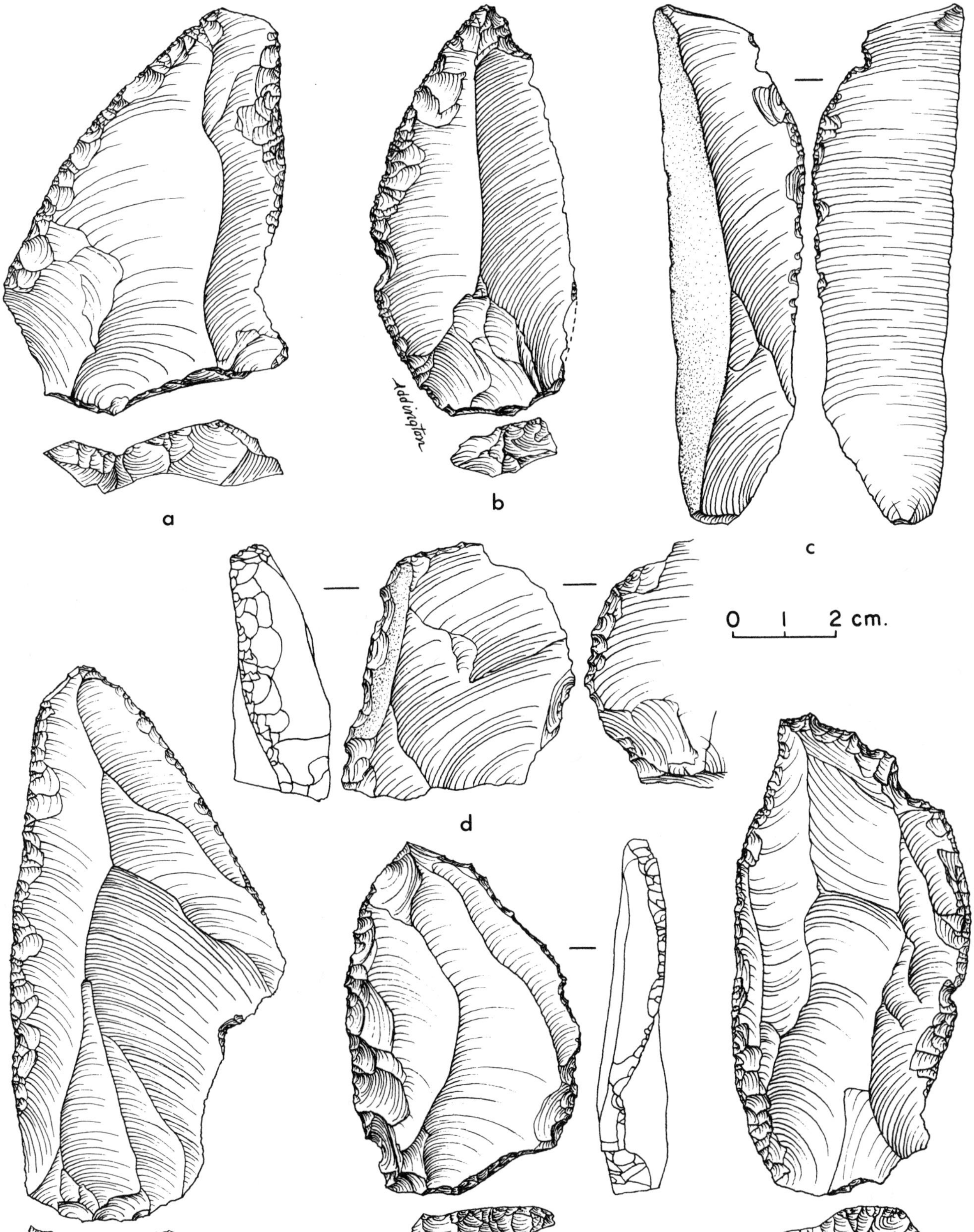

a
b
c
d
e
f
g
0 1 2 cm.
Addington

Figure 23. Massive scraper. Graduated spacing of direction lines in the flake scars of the front view help give illusion of steep-sided piece.

Source: Marks, ed. 1976.

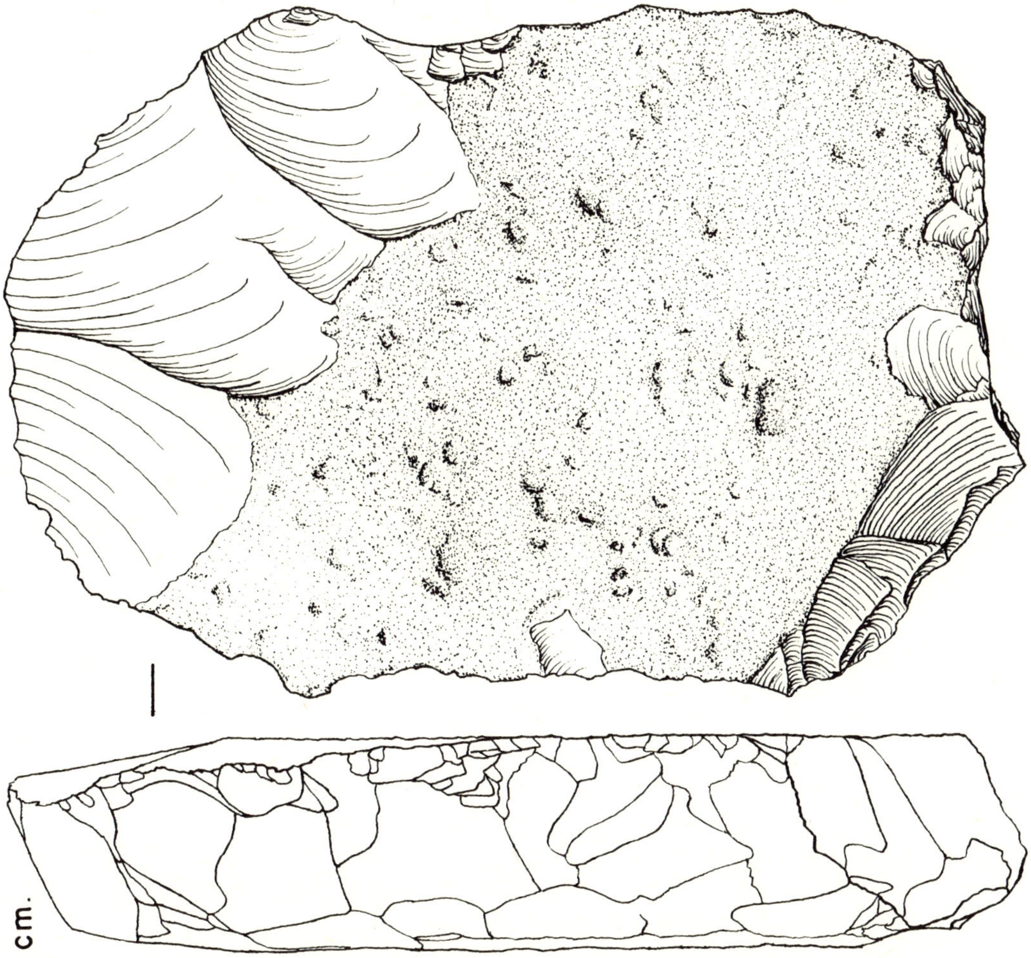
0 1 2 cm.

Figure 24. Various retouched pieces. Note spacing of direction lines on major long flake scars of *d*, *e*, and *i* to depict rippled effect. Note also the treatment of shattered and hinge-fractured scars on *h*. The curvature and placement of direction lines in major left scar of *h* denote that scar was produced before detachment from core; spacing and lengths of the scar's direction lines give effect of scar's "climb" from low left, upward along the side, to descend toward distal end. Older, heavily patinated scars through central part of *j* contrast with newer bilateral and proximal retouch scars.

Sources: *a*, *d*: Marks, ed. 1977; *b*, *h*: Marks, ed. 1983b; *c*, *e*–*g*, *i*, *j*: Wendorf and Schild 1980.

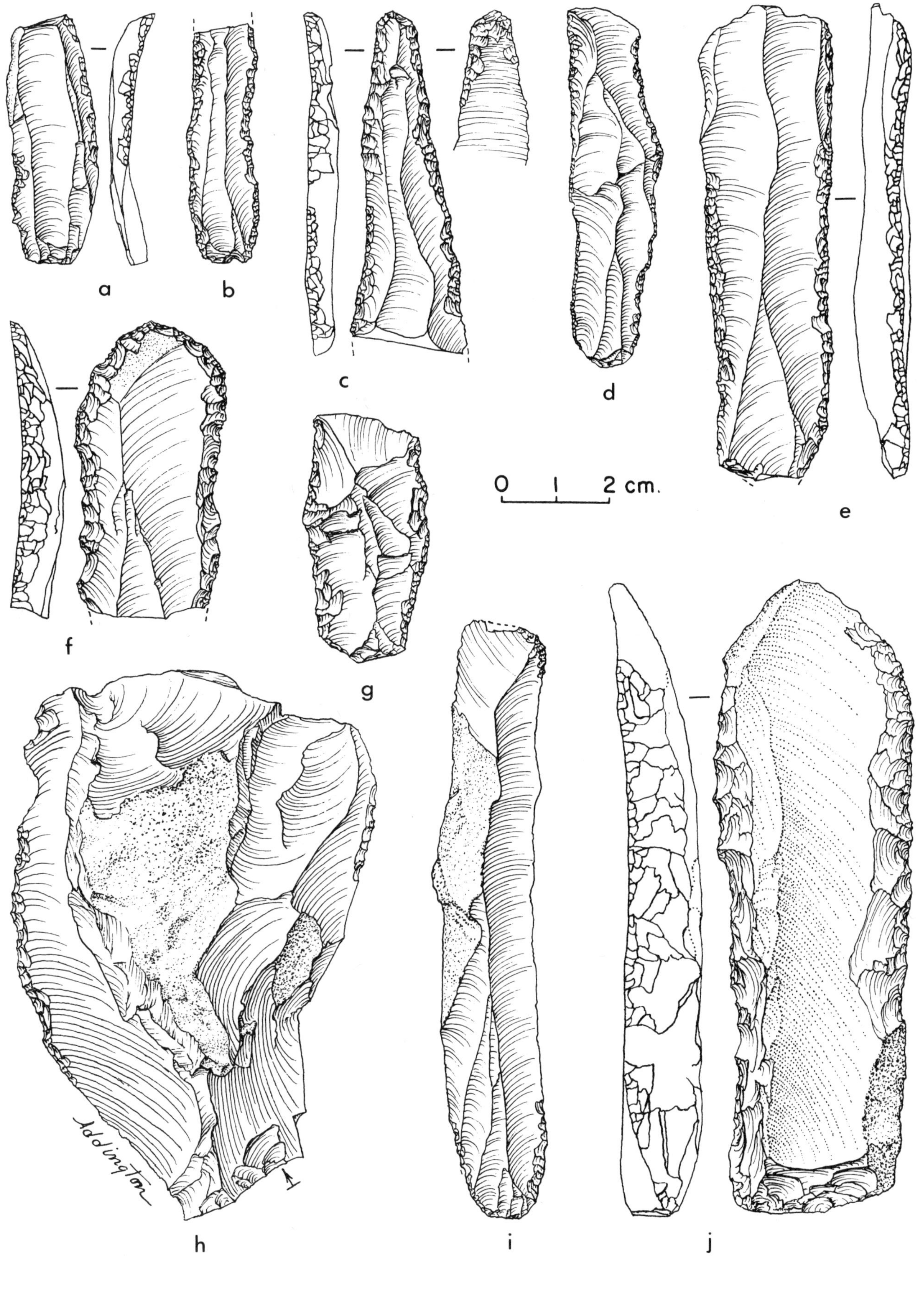
a
b
c
d
e
f
g
h
i
j
0 1 2 cm.
Addington

Figure 25. Examples of retouched blades. *a*, *b*, *f*: rather irregular retouch; *a*, *b*: retouch is visible enough in views of front and back so that edge-only profile is sufficient; *c*, *d*, *e*: rather steep retouch that needs one or more profiles with scar outlines; *f*: an atypical pattern of flake scars that needs explanation by a filled profile.

Sources: *f*: Marks, ed. 1983b; *a*, *b*: Wendorf and Schild 1980; *c*–*e*: Wendorf and Schild, assemblers, and Close, ed. 1984.

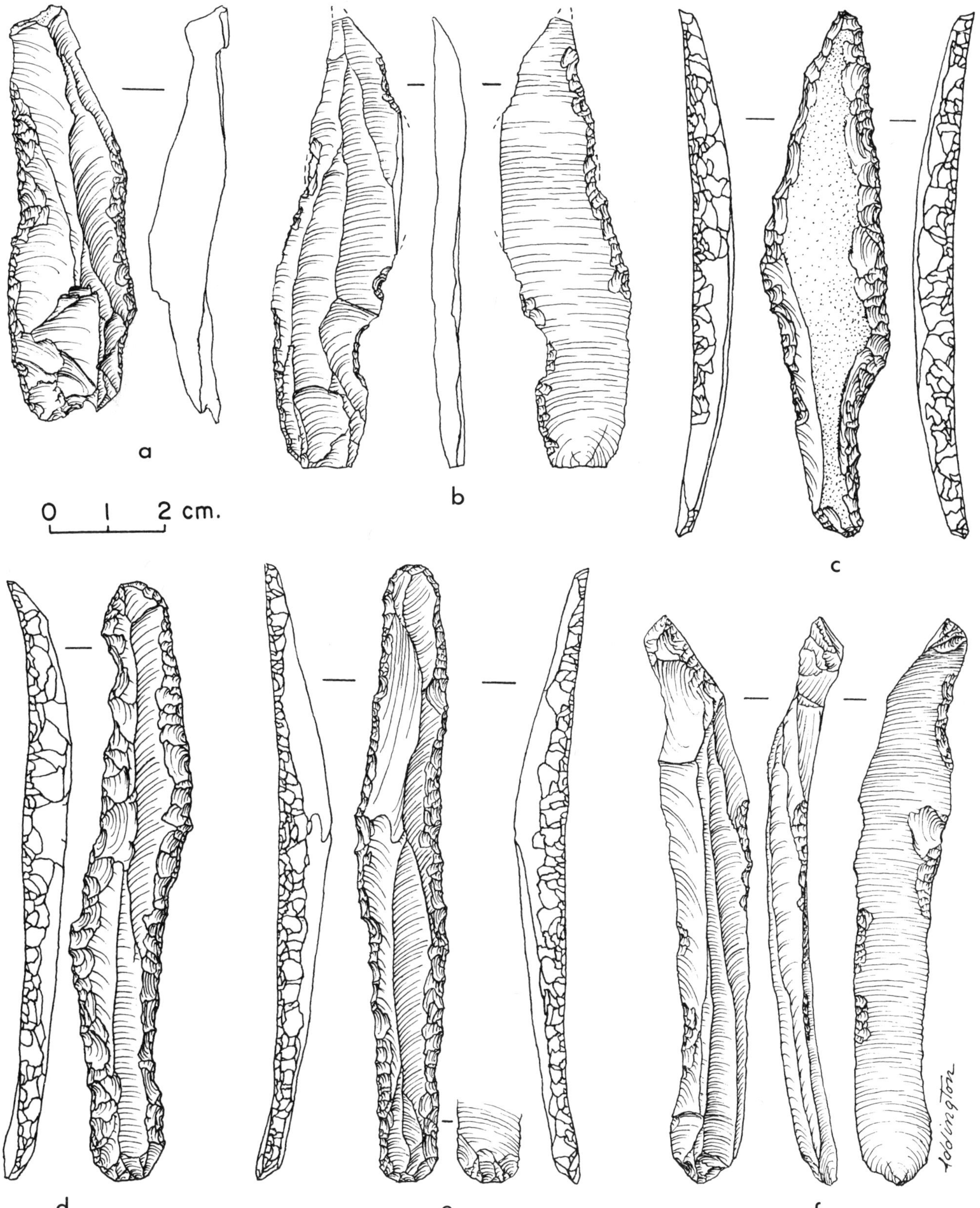
a
b
c
d
e
f
0 1 2 cm.
Addington

Figure 26. Additional examples of retouched pieces. *c*: retouch too fine and delicate to allow direction lines in the scars; *e*, *h*: similar problem of very small retouch scars that would black out if filled with direction lines; *a*, *f*, *j*: examples of off-axis pieces that are straightened to a more balanced position for illustration and barred arrow used to show bulb location and axis of detachment from the core; *k*: retouched piece on a slab with cortex on both front and back. Note that the scar of a fresh surface break on *k* is left blank.

Sources: *a*, *c*, *d*, *f*: Marks, ed. 1976; *b*, *e*, *h*–*j*: Marks, ed. 1977; *g*, *k*, *l*: Wendorf and Schild 1980.

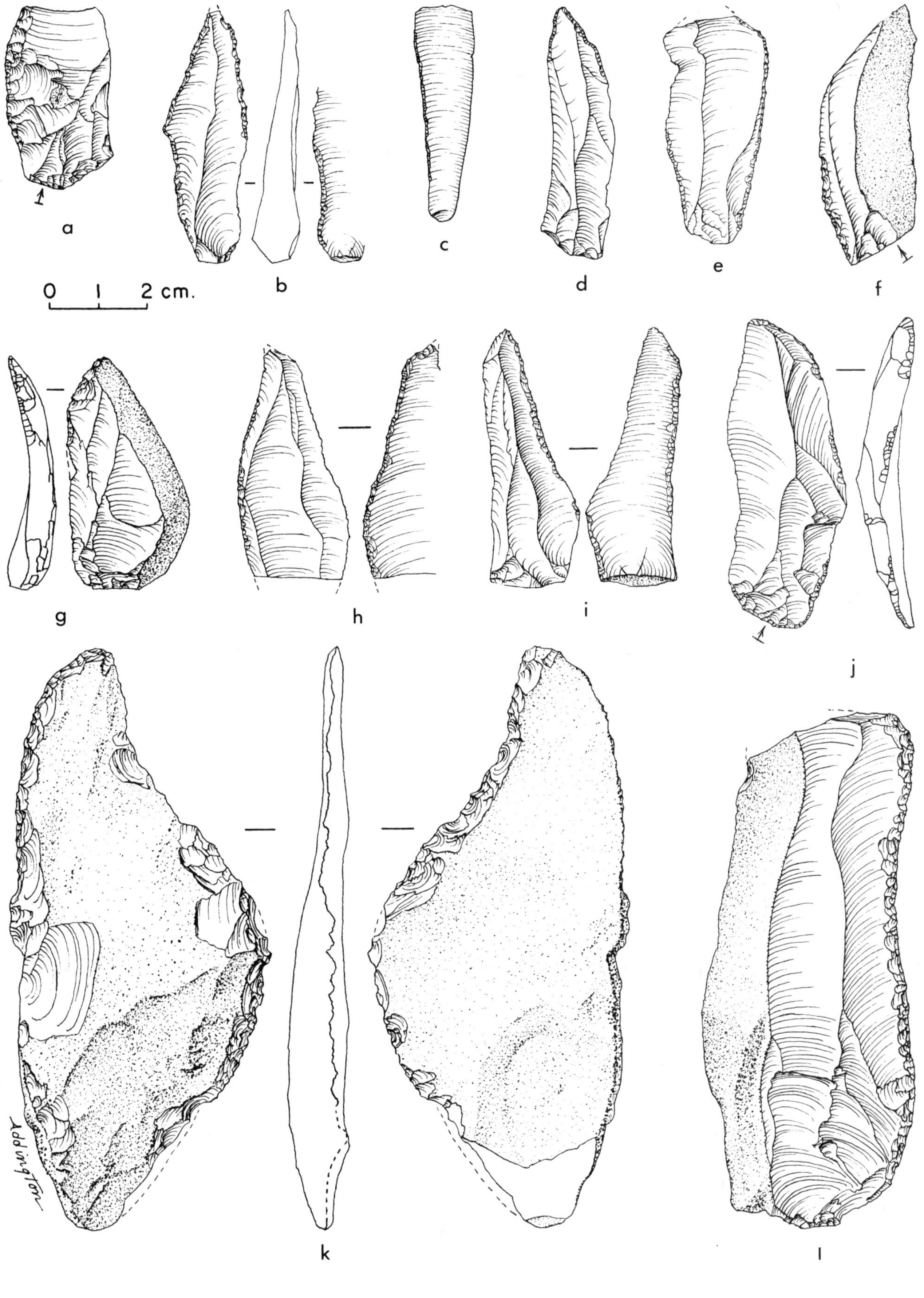
a
b
c
d
e
f
0 1 2 cm.
g
h
i
j
k
l
Addington

Figure 27. Denticulates. *f*, *g*: inversely denticulated; *h*: massive denticulate with a transverse burin, heavily patinated old flake scars, newer denticulation and burin scar, and example of dotted outlining of patinated scars in its empty profile; *i*: massive denticulate with treatment similar to *h* but also has cortex.

Sources: *b*, *d*, *f*-*h*: Marks, ed. 1976; *a*, *c*, *e*, *i*: Marks, ed. 1977.

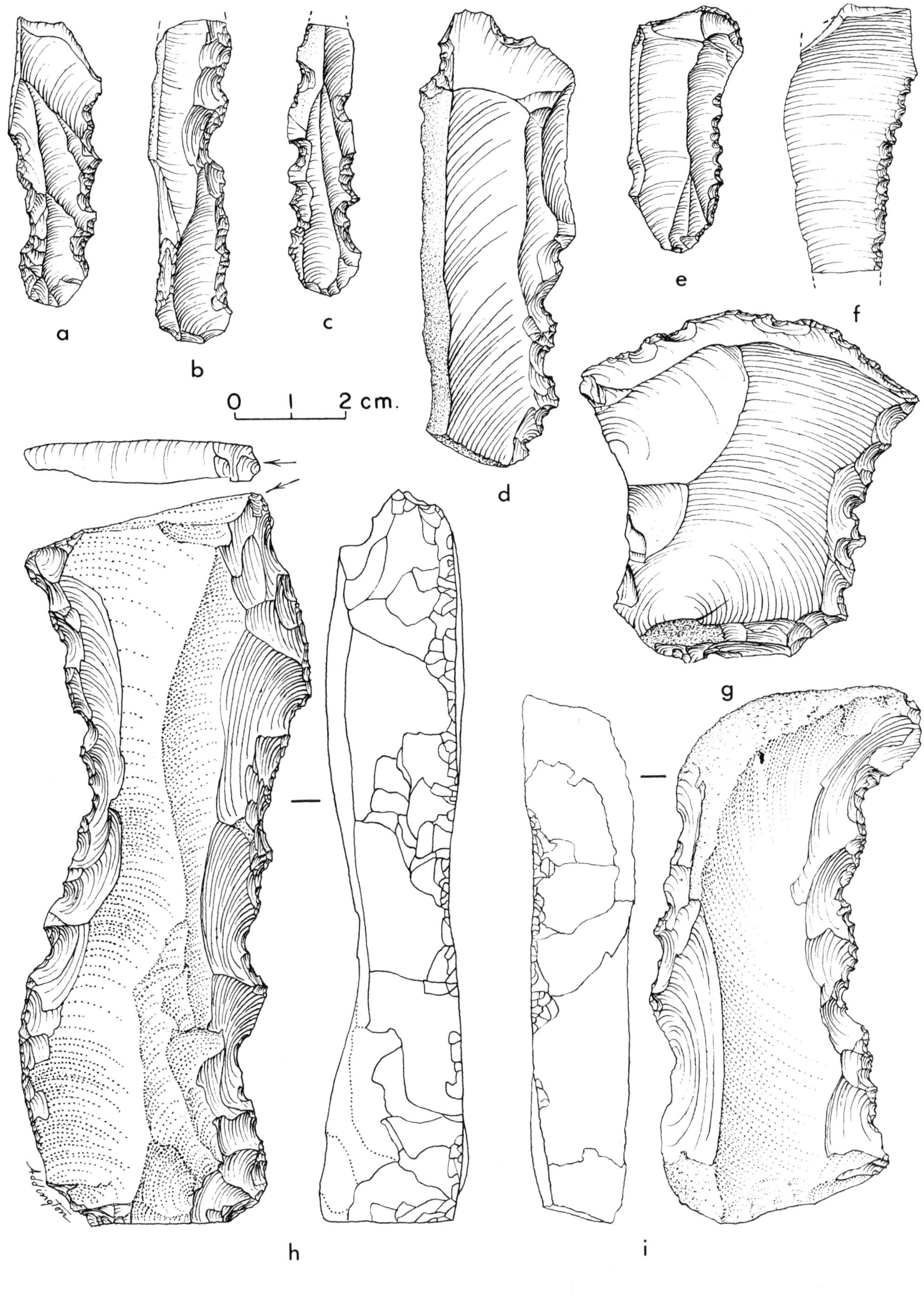
a
b
c
d
e
f
g
h
i
0 1 2 cm.
Addington

Figure 28. Additional examples of denticulated pieces. *e*: such pieces as this unusual denticulated crescent, with complex and elaborate retouch, need careful drawing to retain details of the flaking; *c*: three types of surface (new retouch, older scars, and cortex) on the same piece dictated a filled profile.

Sources: *b*–*d*, *f*: Marks, ed. 1976; *a*, *g*, *h*: Marks, ed. 1983b; *e*: Wendorf and Schild 1980.

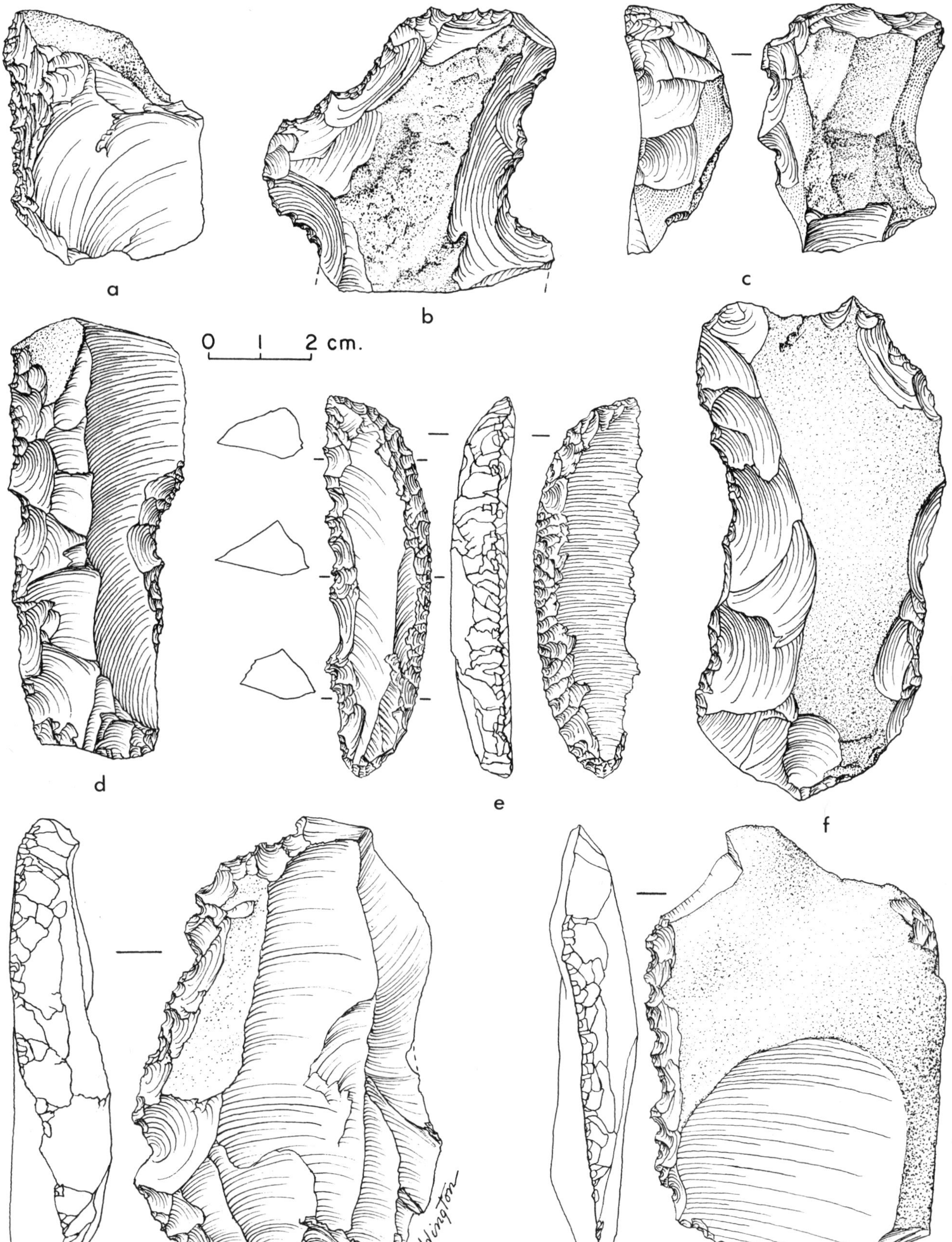
a
b
c
d
0 1 2 cm.
e
f
g
h
Addington

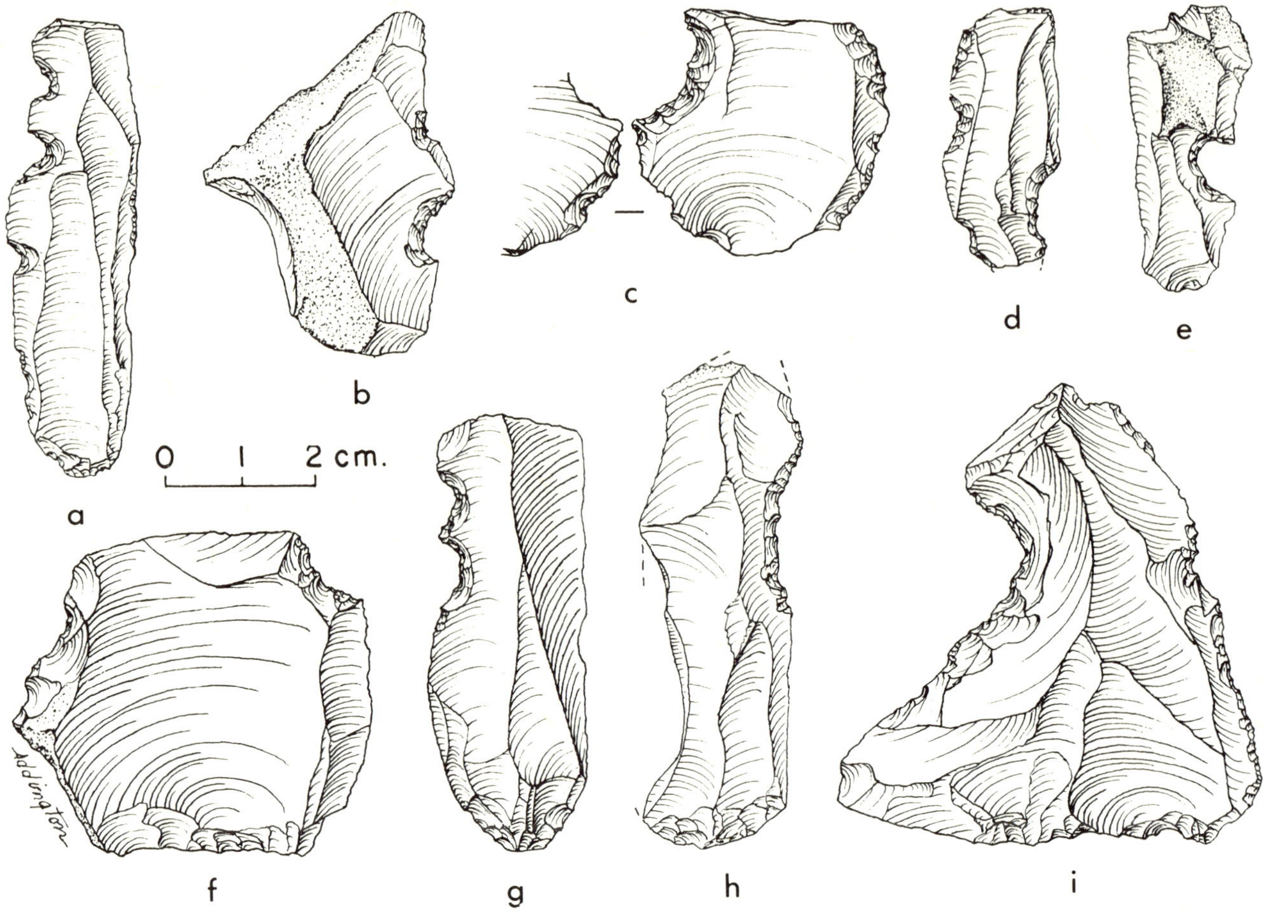

Figure 29. Notched pieces.

Sources: *c*, *f*, *g*, *i*: Marks, ed. 1976; *a*, *d*, *e*: Marks, ed. 1977; *h*: Marks, ed. 1983b; *b*: Wendorf and Schild 1980.

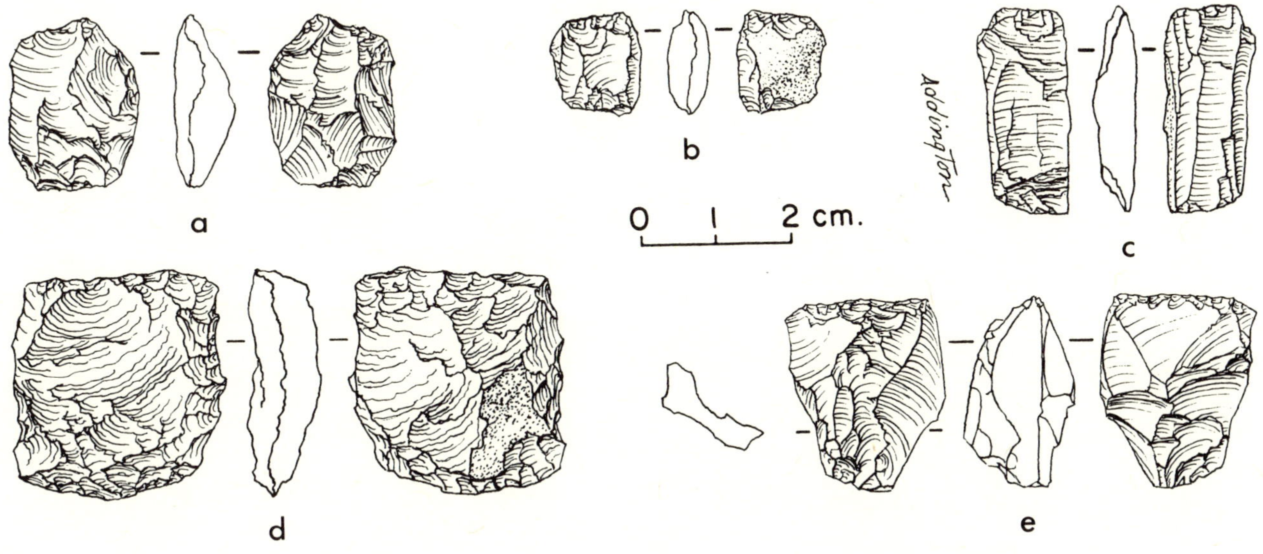

Figure 30. Scaled pieces. *d*: "nervous" direction lines show the raw material of the piece was very rough-fracturing quartz; *e*: atypical twisted form needed explanatory cross section to correct impression from other views that the piece was thick.

Sources: *e*: Wendorf and Schild 1980; *a–d*: Wendorf and Schild, assemblers, and Close, ed. 1984.

Figure 31. Truncated pieces. *a*, *b*: double truncations; *c*, *h*, *m*, *o–t*: distal truncations; *d–g*, *k*: oblique distal truncations; *i*, *j*, *l*: oblique proximal truncations; *n*: proximal truncation; *c*, *h*, *m*: convex; *j–l*, *n*, *o*, *t*: concave. Proximal view of *t* is a butt (or platform) view of its Levallois preparation, not a truncation.

Sources: *b*, *c*, *f*, *m*, *n*, *q–t*: Marks, ed. 1976; *d*, *e*, *g–i*, *o*, *p*: Marks, ed. 1977; *j–l*: Wendorf and Schild 1980.

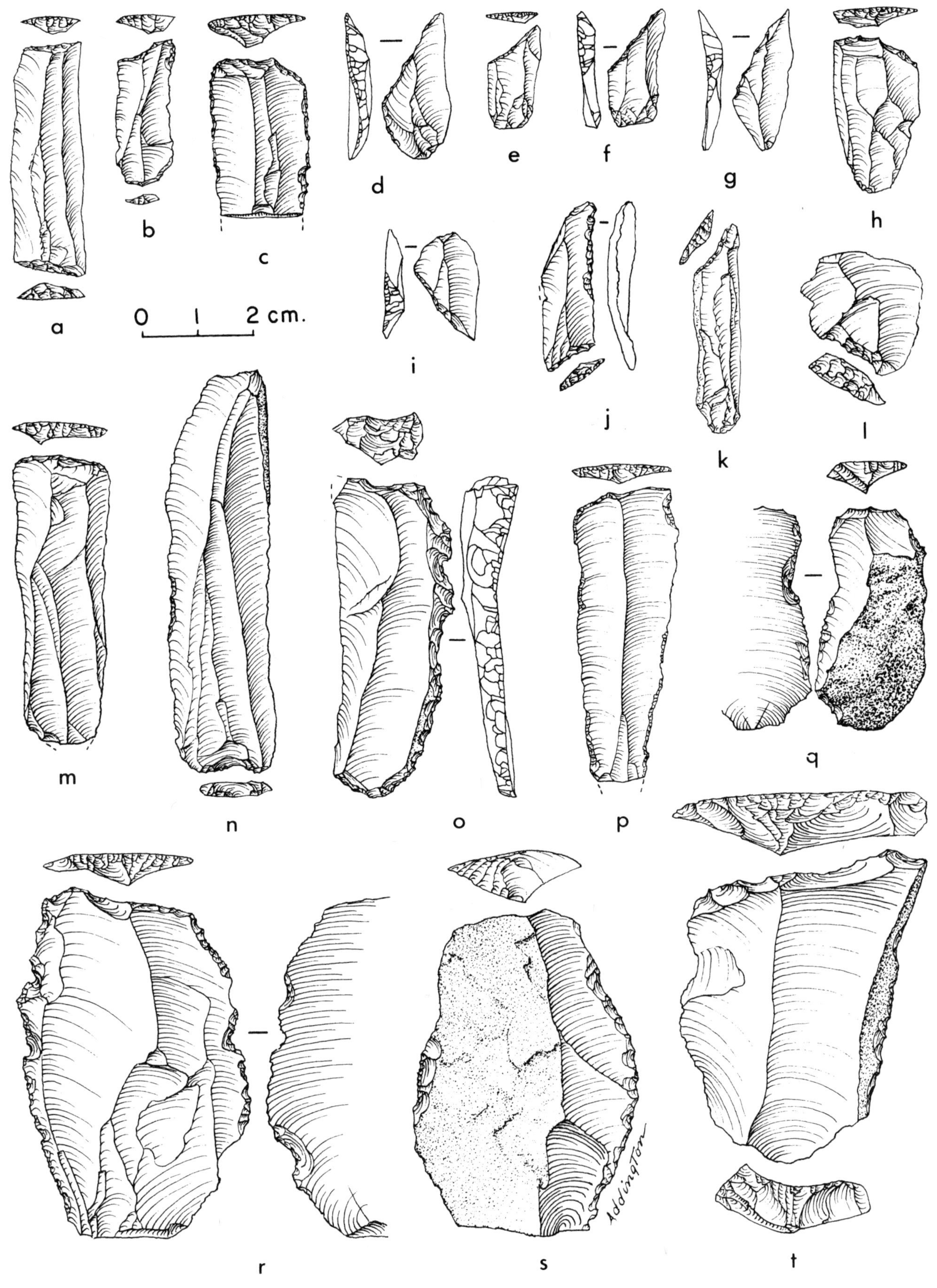

a
b
c
d
e
f
g
h
i
j
k
l
m
n
o
p
q
r
s
t
0 1 2 cm.
Addington

Figure 32. Dihedral burins. *a*, *b*, *g*: multiple dihedral burins; *c*: dihedral angle burin; *d*, *e*, *h*, *i*: dihedral burins; *f*: multiple burin, dihedral and transverse; *j*: burin, multiple mixed (dihedral and on a truncation); *h* has some patinated flake scars with thermal damage.

Sources: *d*, *e*, *i*: Marks, ed. 1976; *c*, *f*, *h*, *j*: Marks, ed. 1983b; *a*, *b*, *g*: Wendorf and Schild, assemblers, and Close, ed. 1984.

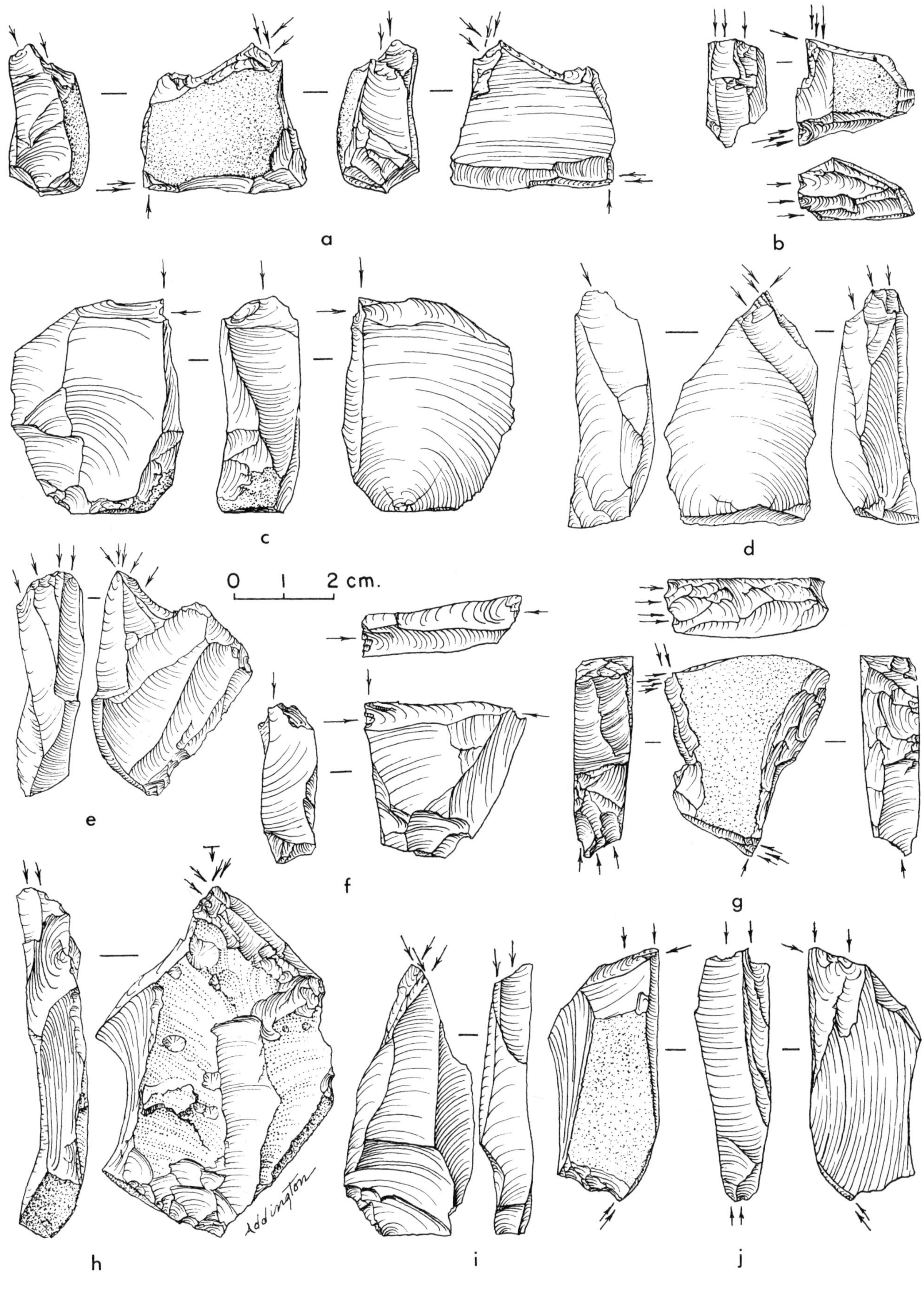
a
b
c
d
e
0 1 2 cm.
f
g
h
i
j
Addington

Figure 33. Burins on snaps and truncations. *a*–*c*: burins on snaps; *d*, *g*: multiple burins on snap; *e*: burins on double truncations; *f*, *h*, *k*, *n*: burins on truncation; *i*: multiple burin on truncation; *j*, *l*: multiple burin on a double truncated piece; *m*, *o*: multifaceted burin on an oblique snap.

Sources: *a*–*d*, *f*, *g*, *i*, *j*, *l*–*o*: Marks, ed. 1976; *e*, *h*, *k*: Marks, ed. 1977.

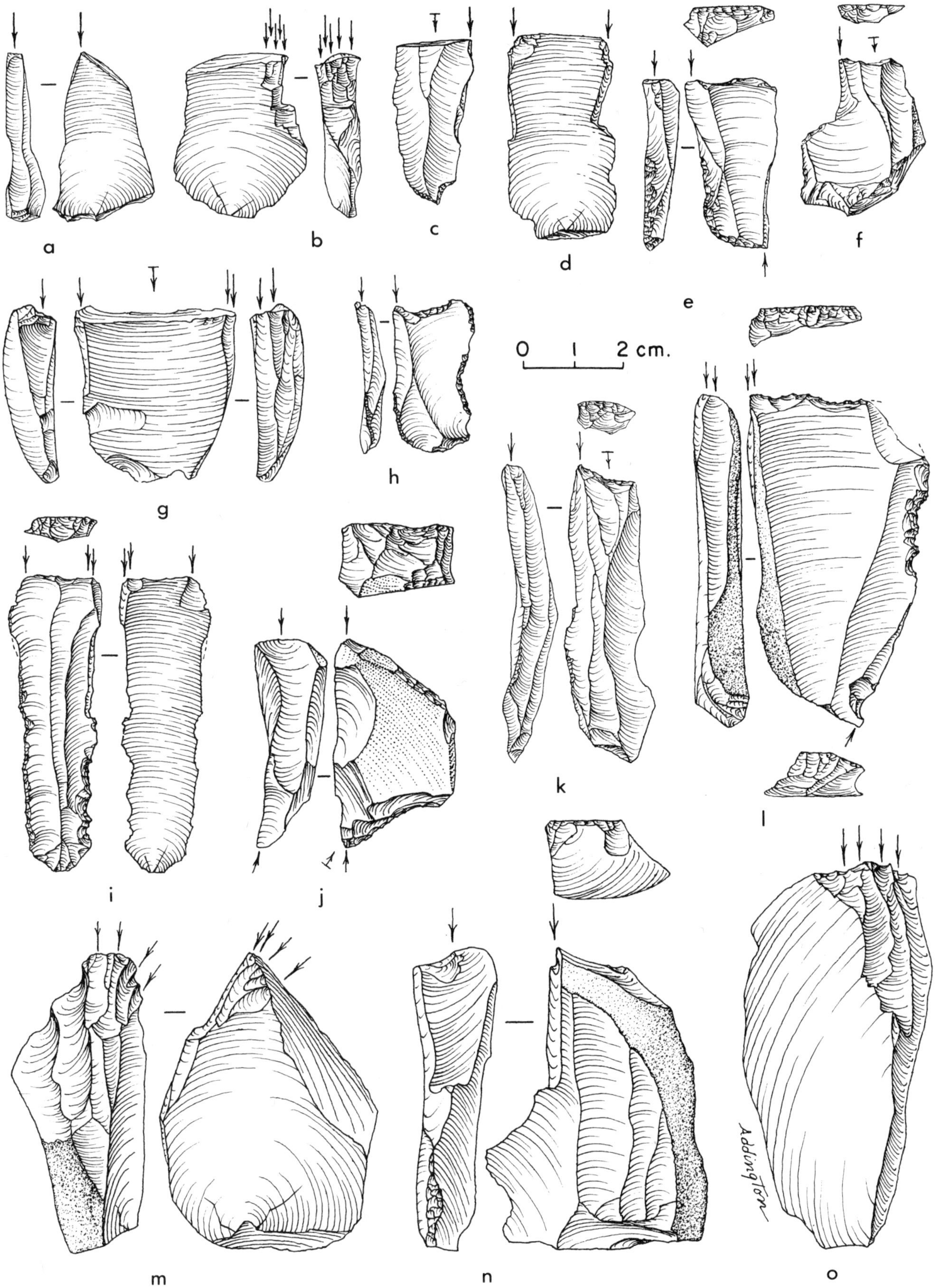
a
b
c
d
e
f
g
h
0 1 2 cm.
i
j
k
l
m
n
o
Addington

Figure 34. Additional types of burins, partly reconstructed burins, and burin spalls. *a*, *d*: burins on preexisting surface; *b*, *e*: busquoid burins; *c*: nucleiform burin with burin spall; *f*: overpassed burin spall; *g*: partly reconstructed burins on a truncation; *h*: carinated burin.

Sources: *a*, *c*, *g*: Marks, ed. 1983b; *b*, *d*, *e*, *h*: Marks, ed. 1976; *f*: Wendorf and Schild, assemblers, and Close, ed. 1984.

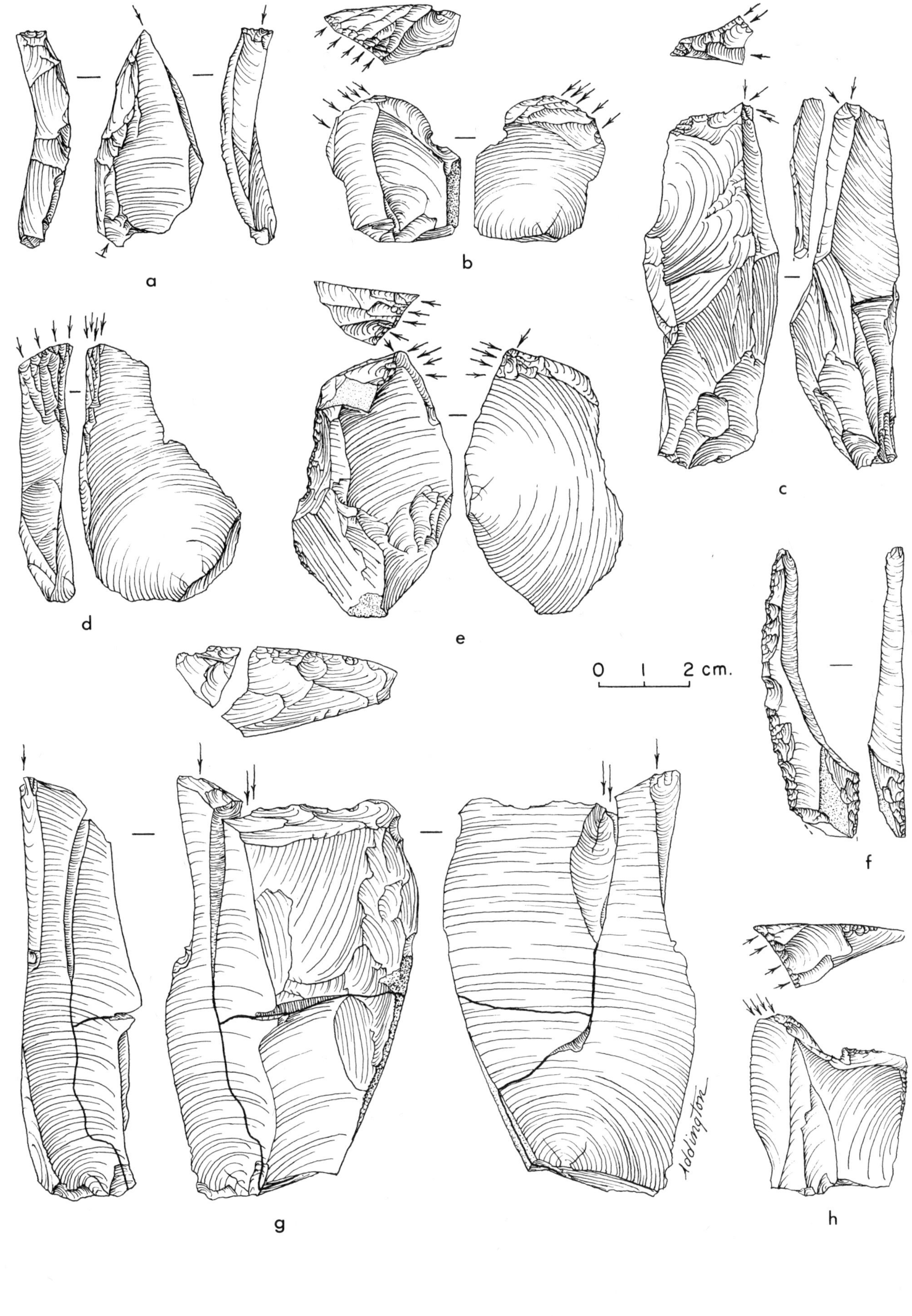
a
b
c
d
e
f
g
h
0 1 2 cm.

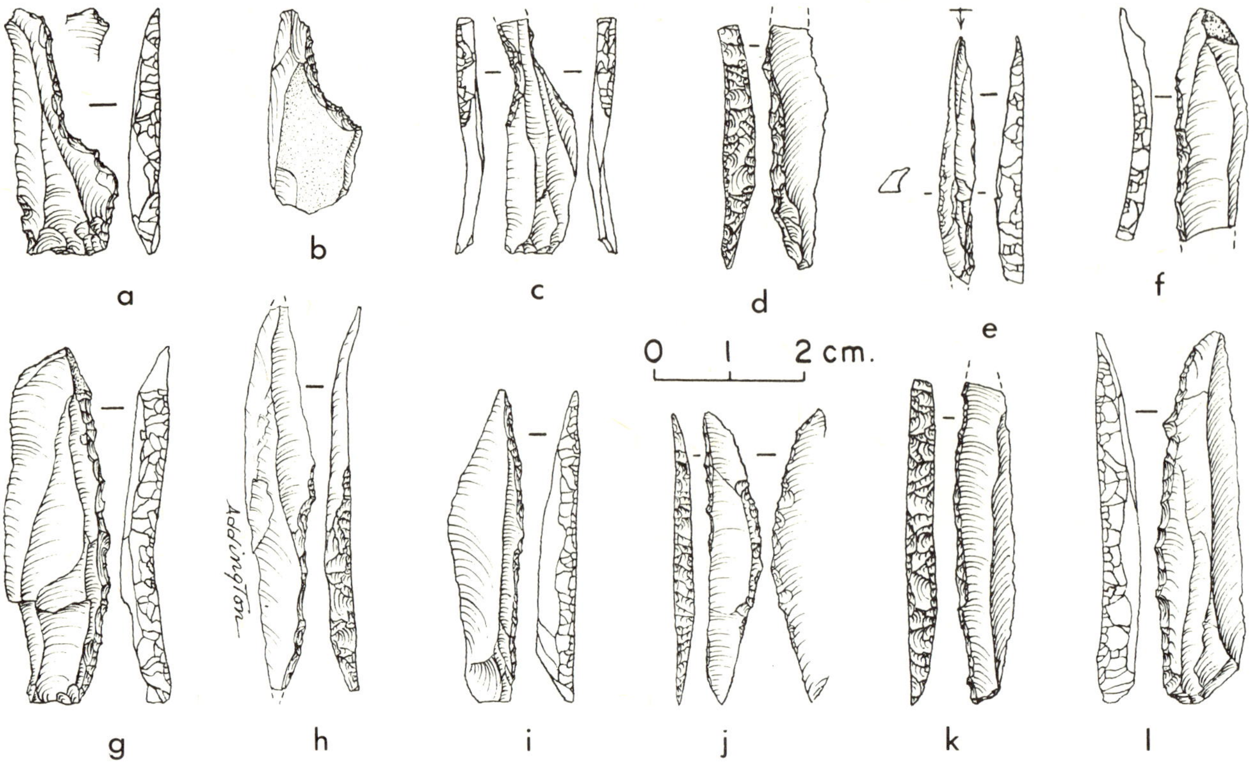

Figure 35. Backed blades and bladelets. *a*, *b*, *h*: shouldered pieces; *c*: strangled blade; *d*, *e*, *g*, *i–l*: backed blades and bladelets; *f*: partially backed bladelet. Note that *d*, *h*, *j*, *k* have bipolar or bidirectional backing that requires filled profiles; *a* has a partial view of the bit of inverse retouch on its ventral surface; *e* needed the transverse cross section to show the extreme undercutting of its backing; *j* has a full-length partial view of the pronounced use retouch along its cutting edge, although a complete ventral view, instead of a partial view, would have been more appropriate for such a narrow bladelet.

Sources: *a*, *b*, *d*, *g*, *i–l*: Marks, ed. 1977; *c*, *h*: Marks, ed. 1983; *e*, *f*: Wendorf and Schild 1980.

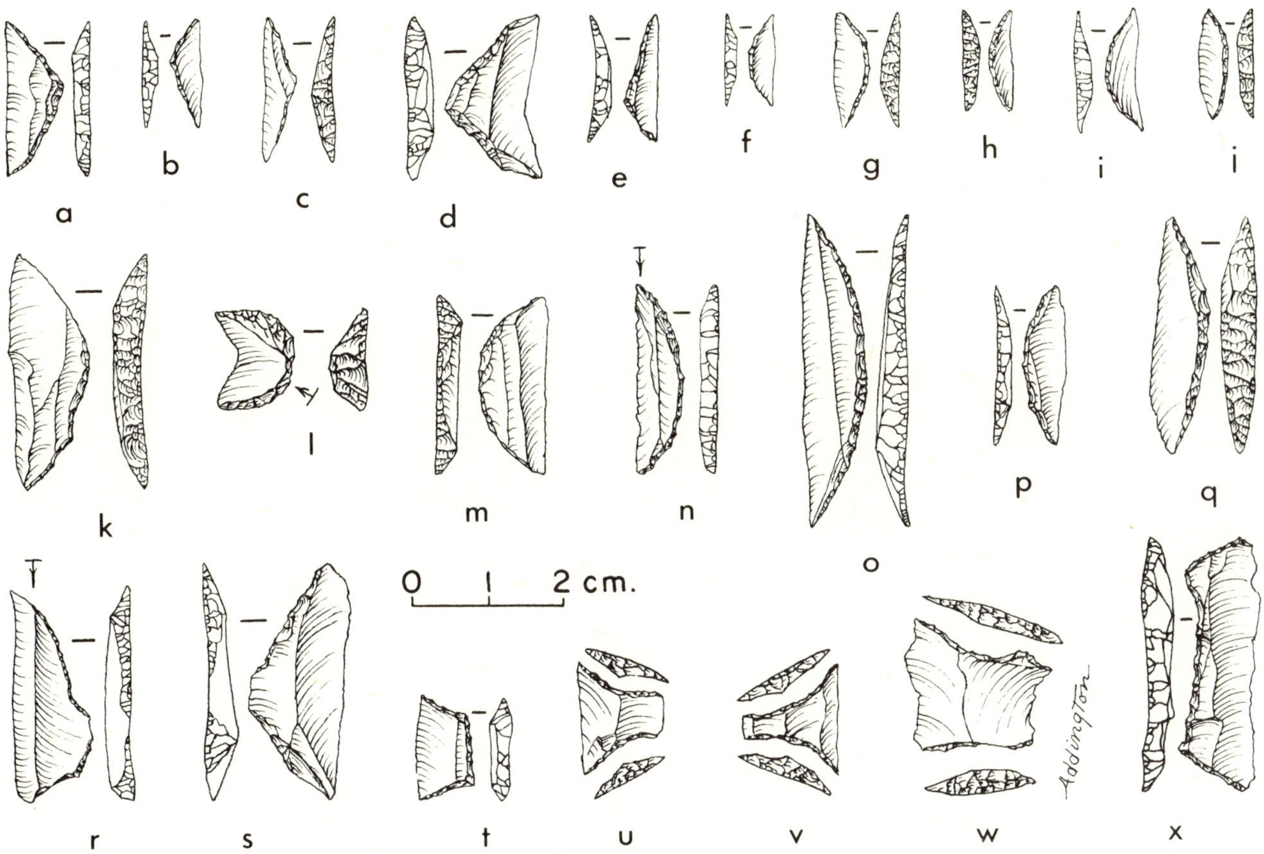

Figure 36. Geometrics. *a-d*: triangles; *e*: scalene triangle; *f-q*: lunates; *r-x*: trapezes. Note that pieces with bipolar (bidirectional) backing require filled profiles; empty profiles mean the backing is obverse (force applied in one direction, from the ventral toward the dorsal surface).

Sources: *t*, *x*: Marks, ed. 1976; *a-d*, *f-k*, *m-s*: Marks, ed. 1977; *e*, *l*, *u-w*: Wendorf and Schild 1980.

Figure 37. Microburins and allied products of the microburin technique. *a–c*: proximal microburins; *d–g*: distal microburins; *h*, *i*: *piquants-trièdres*; *j*, *k*: Krukowski microburins; *m–p*: La Mouillah points; *l*: pieces found to fit together (distal: Krukowski mocroburin; proximal: La Mouillah point). *b–g*, *k*: examples of microburin scars with bulbar treatment; *a*, *h–j*, *n–p*: examples of microburin scars with flake scar treatment; *l*: example of each on the two surfaces that fit together.

Sources: *c*, *f*: Marks, ed. 1976; *a*, *b*, *d*, *e*, *g*, *j*: Marks, ed. 1977; *h*, *i*, *k*, *m–p*: Wendorf and Schild 1980; *l*: Wendorf and Schild, assemblers, and Close, ed. 1984.

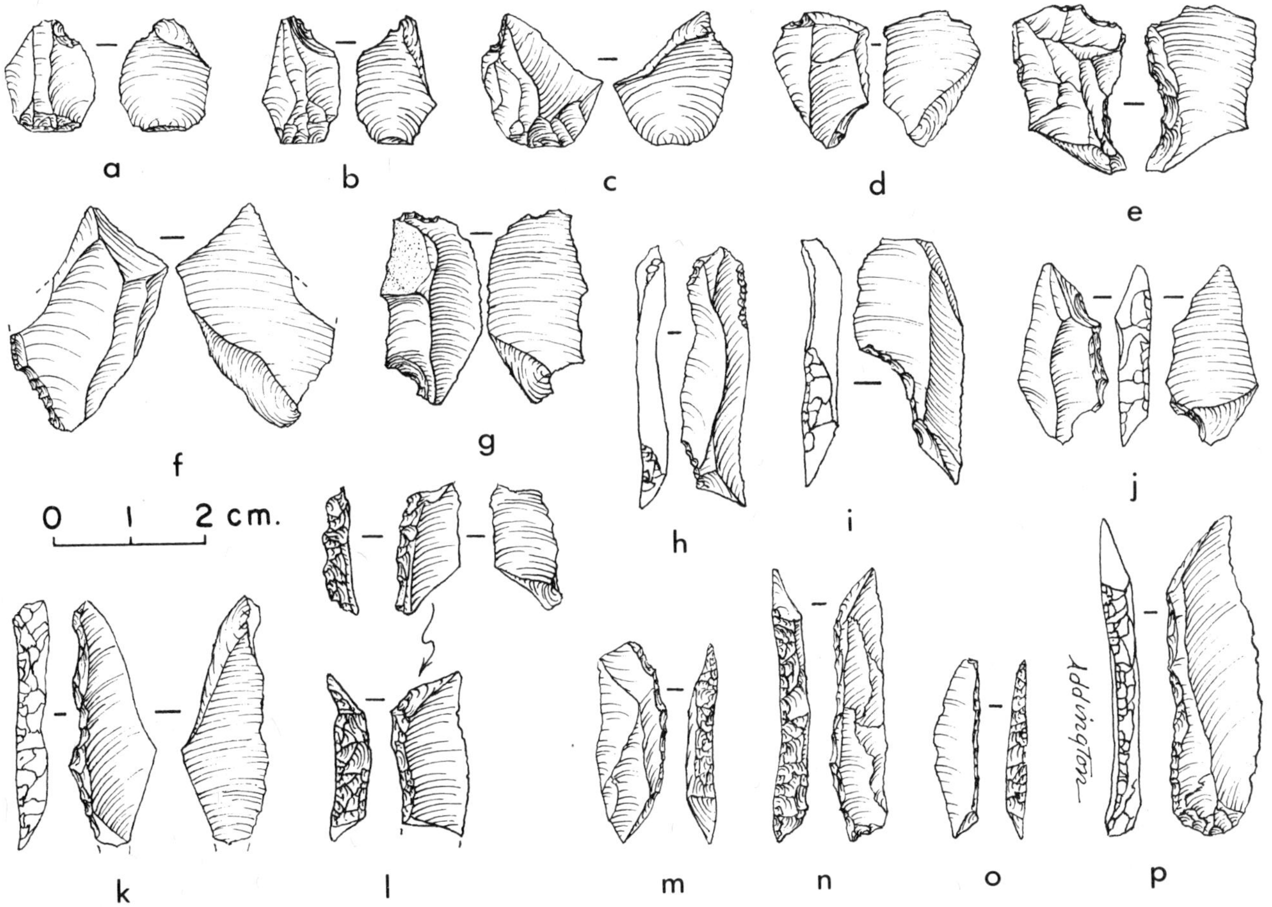
a
b
c
d
e
f
g
h
i
j
0 1 2 cm.
k
l
m
n
o
p
Addington

Figure 38. Perforators (retouched, backed, and double backed) and perforators on burin spalls. *a–f*: double backed; *g*, *h*: backed; *i*: retouched; *j–m*: perforators on burin spalls. *b*, *e*, *f*, *h*, *j*, *l*: inversely retouched; *e*, *f*, *k*, *l*: bidirectional backing; *c*: proximal truncated. Rows of dots and broken flake scar outlines of distal portion of *g* indicate severe wear; cross sections needed to clarify shapes of *f* and *k*.

Sources: *b*, *c*: Marks, ed. 1976; *a*, *d*: Marks, ed. 1977; *h*: Marks, ed. 1983b; *e*, *f*, *i*: Wendorf and Schild 1980; *g*, *j*, *k*, *m*: Wendorf and Schild, assemblers, and Close, ed. 1984.

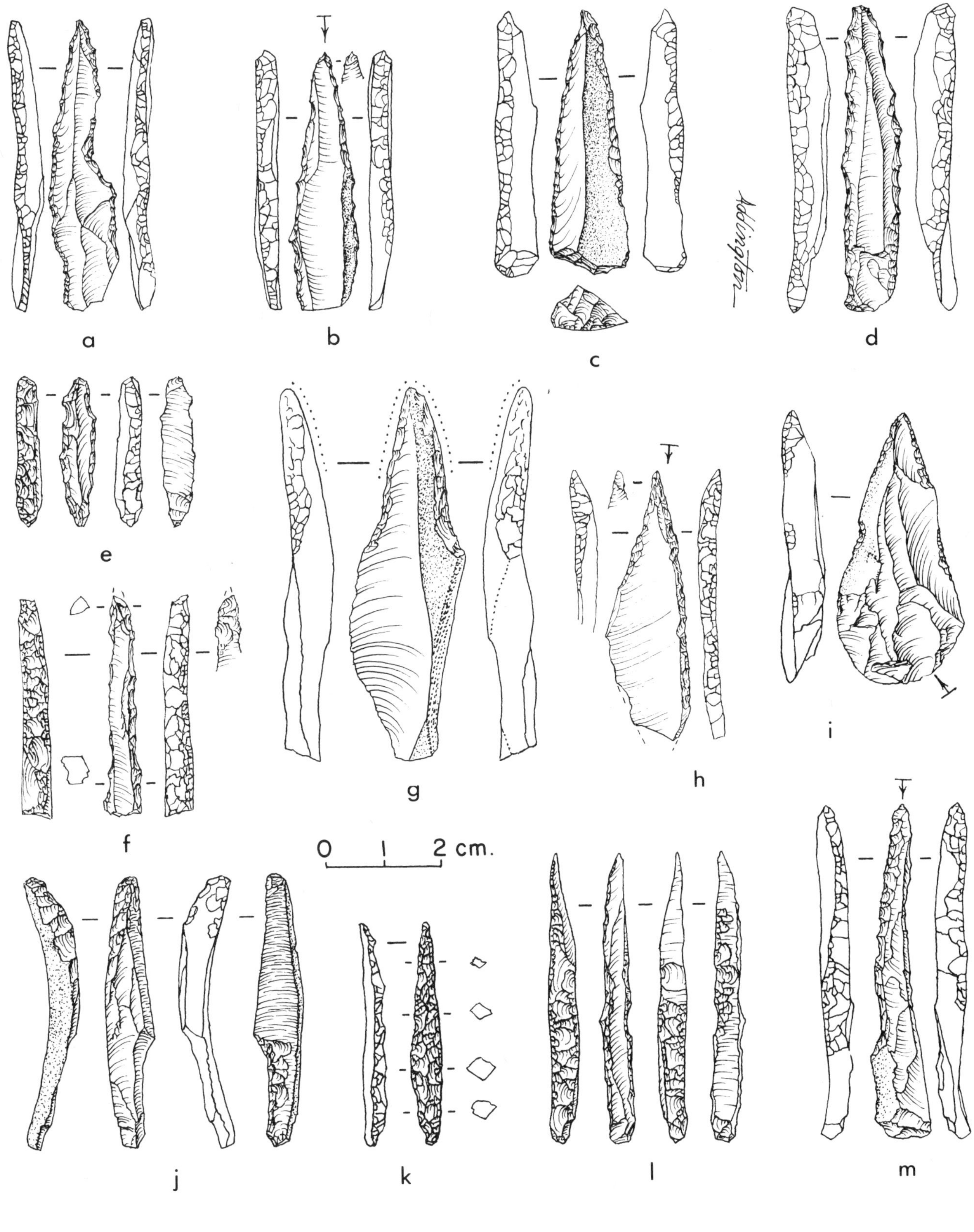
a
b
c
d
e
f
g
h
i
0 1 2 cm.
j
k
l
m

Figure 39. Other varieties of perforators. *e*, *l*-*o*: examples of "tucked" views that save space. Note, however, that they retain horizontal alignment with primary views.

Sources: *f*, *m*: Marks, ed. 1976; *b*, *d*, *e*: Marks, ed. 1977; *a*, *c*, *g*-*l*, *n*, *o*: Wendorf and Schild 1980.

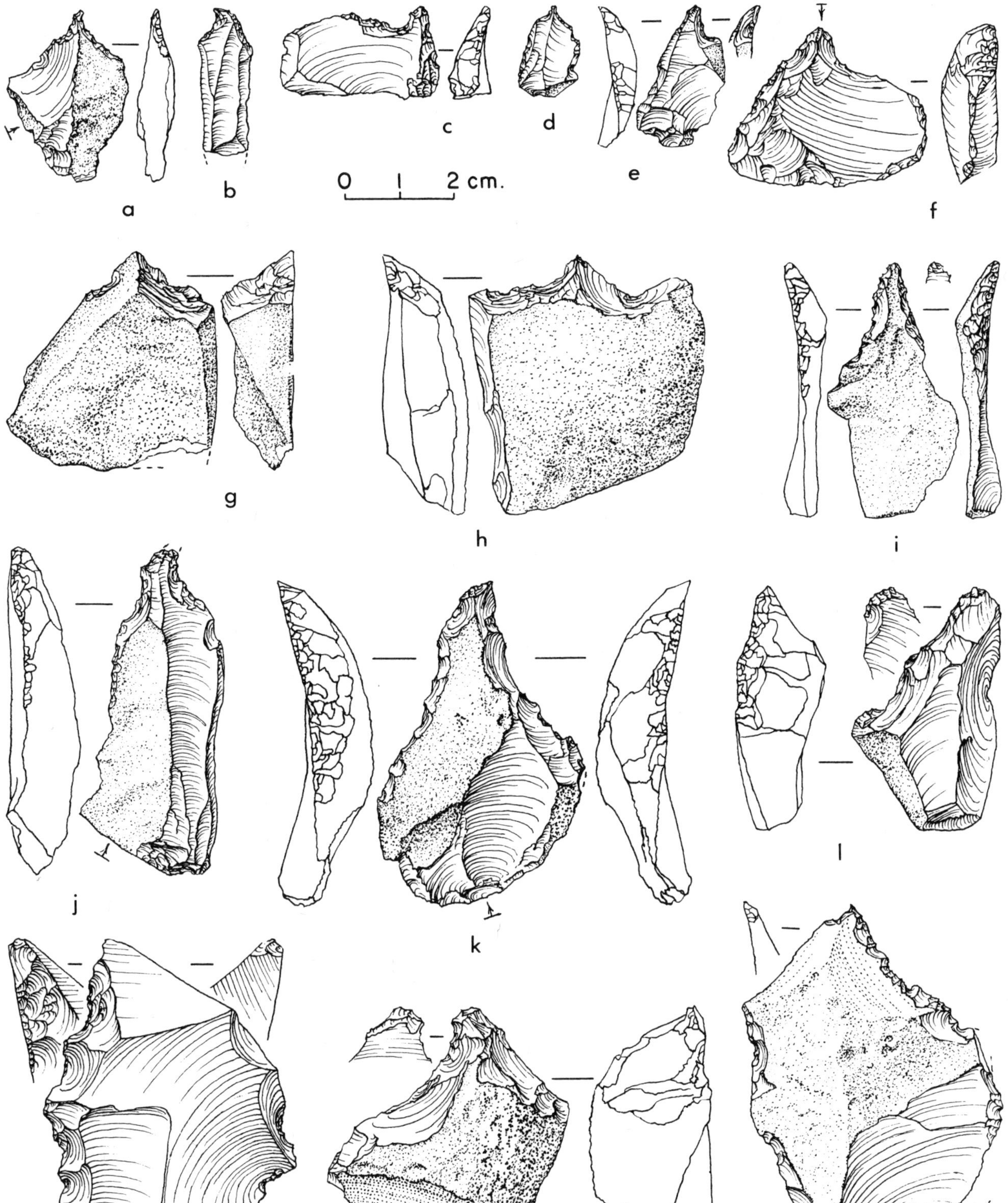
a
b
c
d
e
f
0 1 2 cm.
g
h
i
j
k
l
m
n
o
Addington

Figure 40. Perforators and other pointed pieces. *a*: double perforator with extreme wear of distal end, showing almost obliterated retouch scars and longitudinal strips of pronounced multidirectional striations; *b*: borer with tucked partial view of the ventral tip; *c*: perforator with complex bidirectional retouch; *d*: simple perforator on a flake; *e*: double backed pointed piece with inverse retouched tip; *f*: pick-like piece.

Sources: *b*, *e*, *f*: Marks, ed. 1976; *a*, *c*, *d*: Wendorf and Schild 1980.

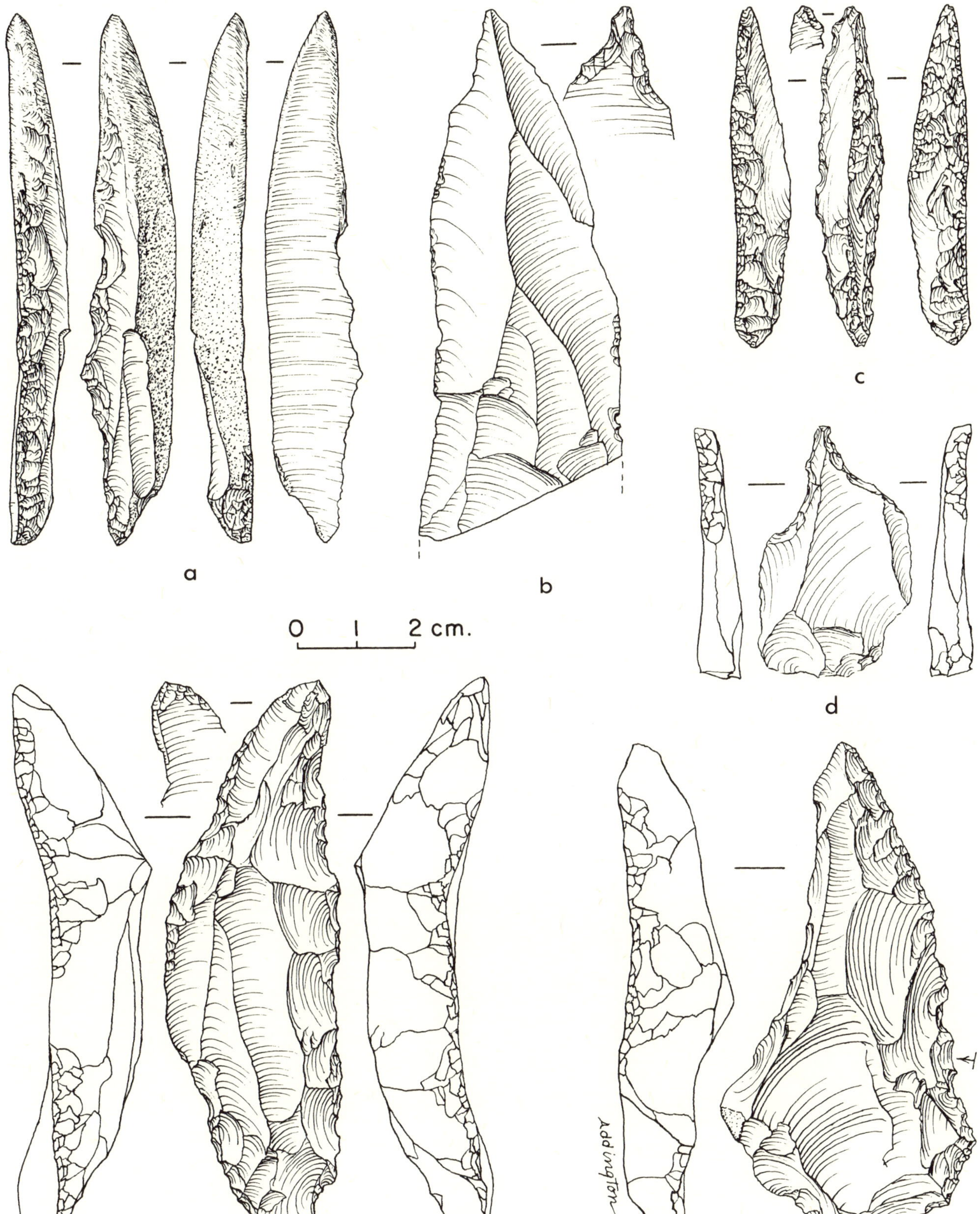
a
b
c
d
e
f
0 1 2 cm.
Addington

Figure 41. Small tanged, stemmed, and serrated points. *a*, *f*, *p*: Neolithic points; *b*: stemmed point; *c*, *k-o*: tanged points; *d*: Washita; *e*: Reed; *g*, *h*: Perdiz; *i*: Scallorn; *j*: Alba.

Sources: *d*, *e*: Henry 1978a; *g-j*: Henry et al. 1980; *c*, *k-o*: Marks, ed. 1976; *a*, *f*, *p*: Marks, ed. 1977; *b*: Wendorf and Schild 1980.

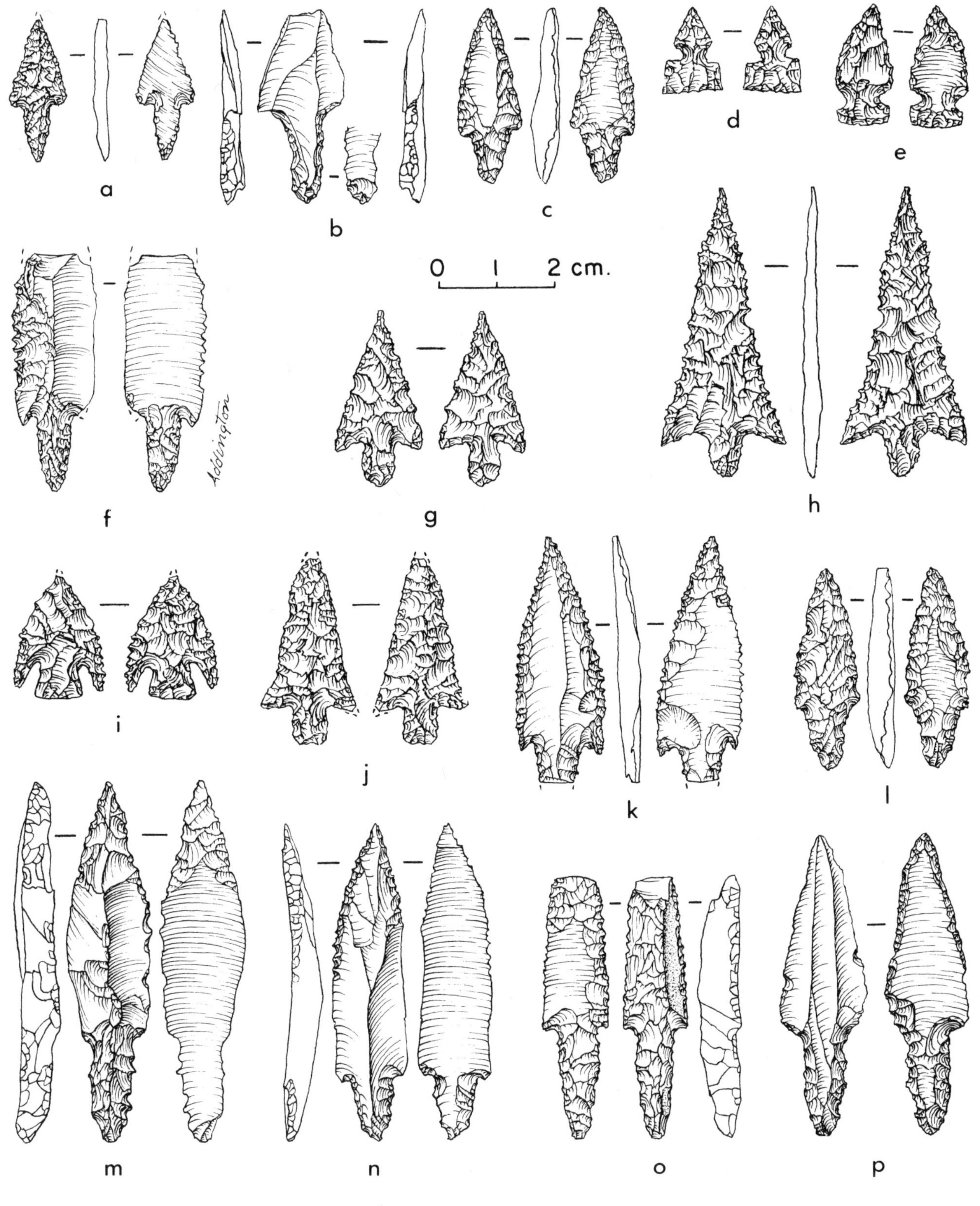
a
b
c
d
e
0 1 2 cm.
f
Addington
g
h
i
j
k
l
m
n
o
p

Figure 42. Arrow and dart points. *a*, *f*: Ellis; *b*, *h*: lanceolate; *c*: Fresno; *d*: Ensor; *e*: Darl; *g*: contracting stem point; *i*: Yarbrough. Note indication of ventral surfaces on *a* and *c*; also use of cross sections with *e* and *f*.

Sources: *b*, *c*, *g*–*i*: Henry 1978a; *a*, *d*–*f*: Henry et al. 1980.

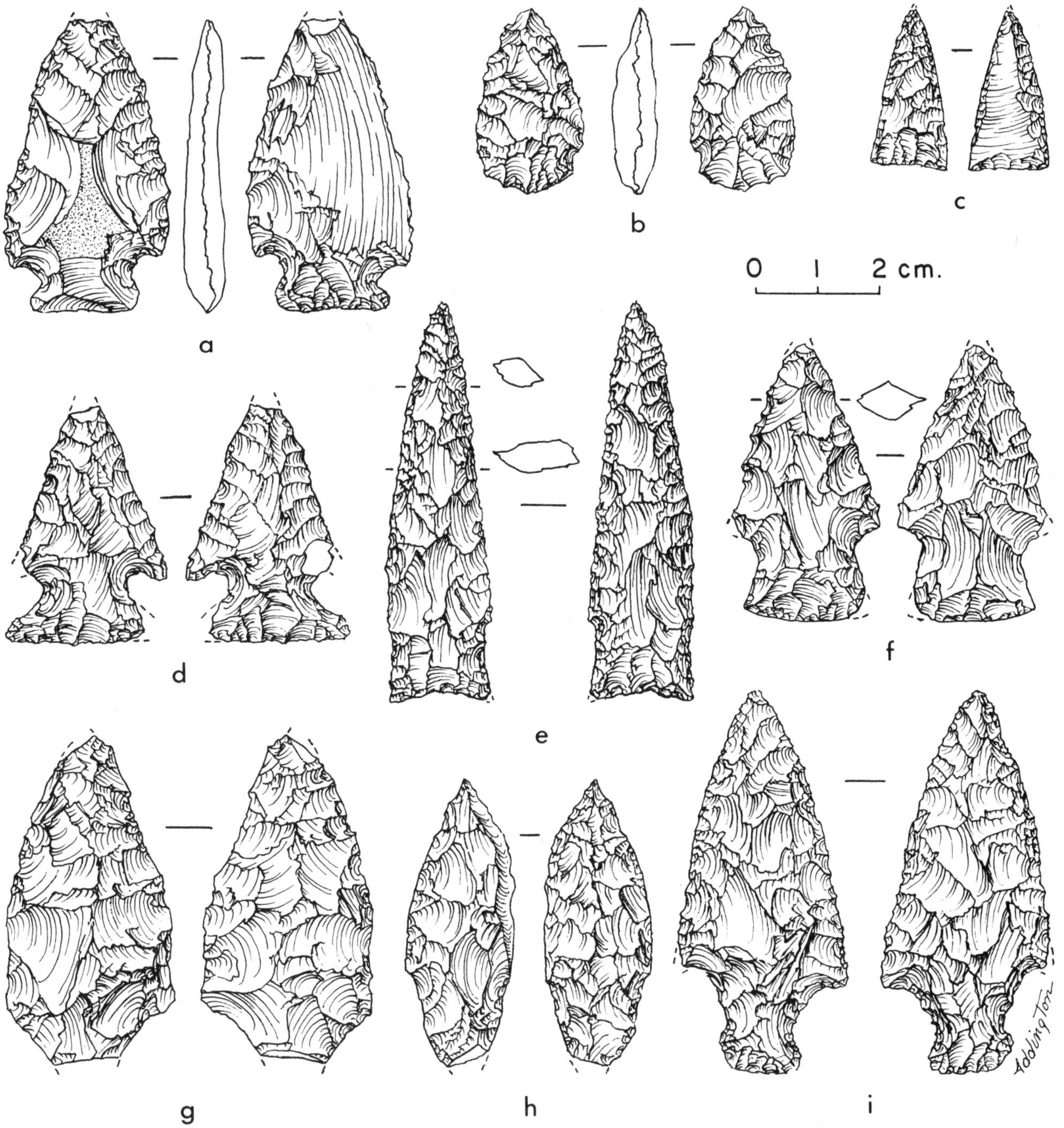
a
b
c
0 1 2 cm.
d
e
f
g
h
i
Addington

Figure 43. Projectile points on blades and bladelets. *a*, *b*: el Wad points; *c*: untanged point; *d*: backed point; *e*: Ounan point; *f*: Harif point; *g*, *h*: Levallois points; *i*, *k*: Emireh points; *j*: notched Levallois point. Note that the use of edge-only profiles with *i* and *k* clarifies the characteristic thinning of the butts.

Sources: *a*–*c*, *j*: Mark, ed. 1976; *e*, *f*: Marks, ed. 1977; *g*–*i*, *k*: Marks, ed. 1983b.

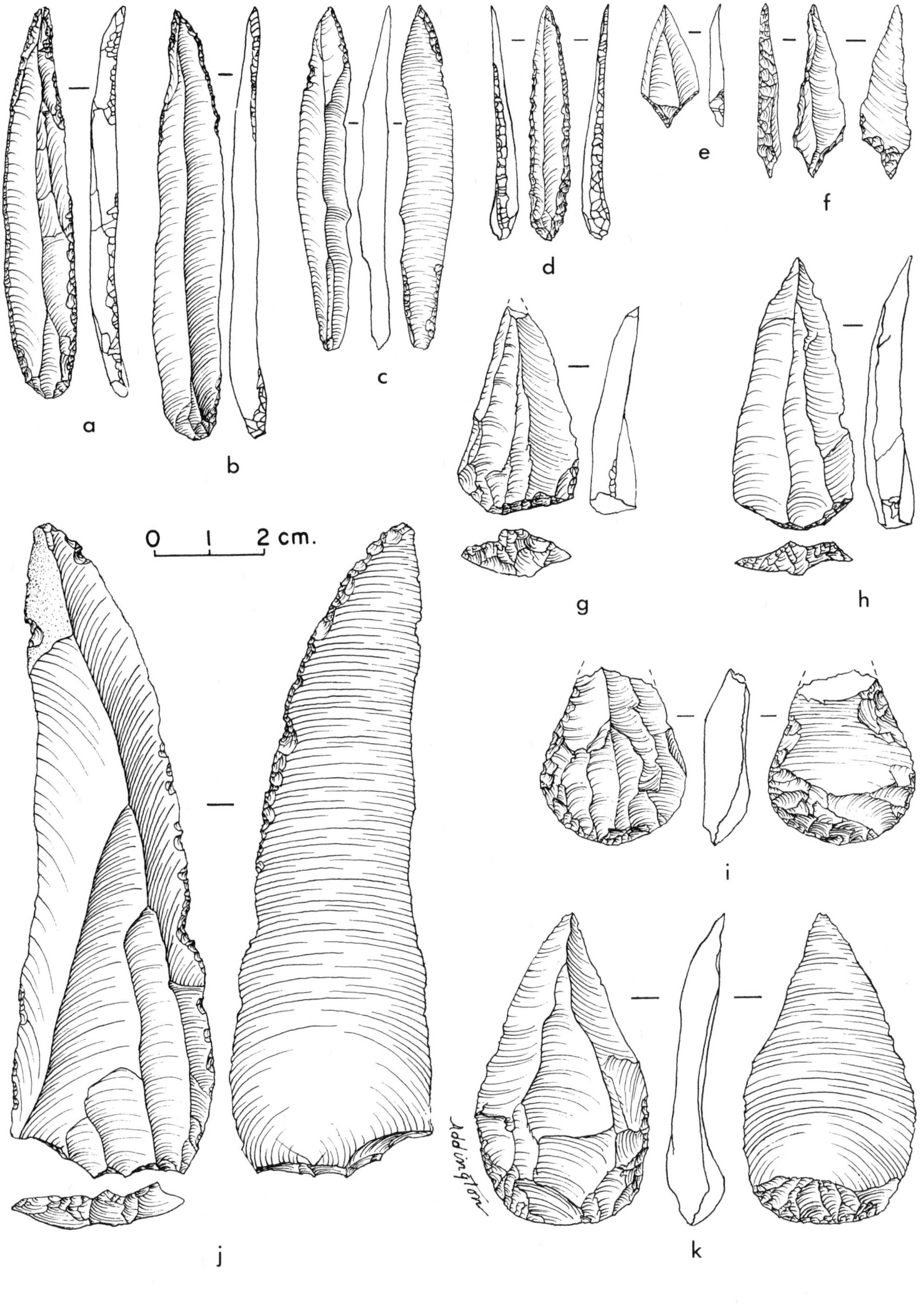
a
b
c
d
e
f
g
h
0 1 2 cm.
i
j
Addington
k

Figure 44. Incurvate and lanceolate bifacial pieces. *a*, *b*, *d*: lanceolate; *c*: incurvate bifacial piece with beveled shape clearly emphasized by the two transverse sections; *a*: center portion of left view is large area of thermal damage. Note that the small runners or ladders, drawn as part of many flake scar outlines, particularly in *c* and *d*, show the overlapping of scars and therefore the order in which removals were made, a part of the pattern of manufacture.

Source: *a*–*d*: Henry et al. 1980.

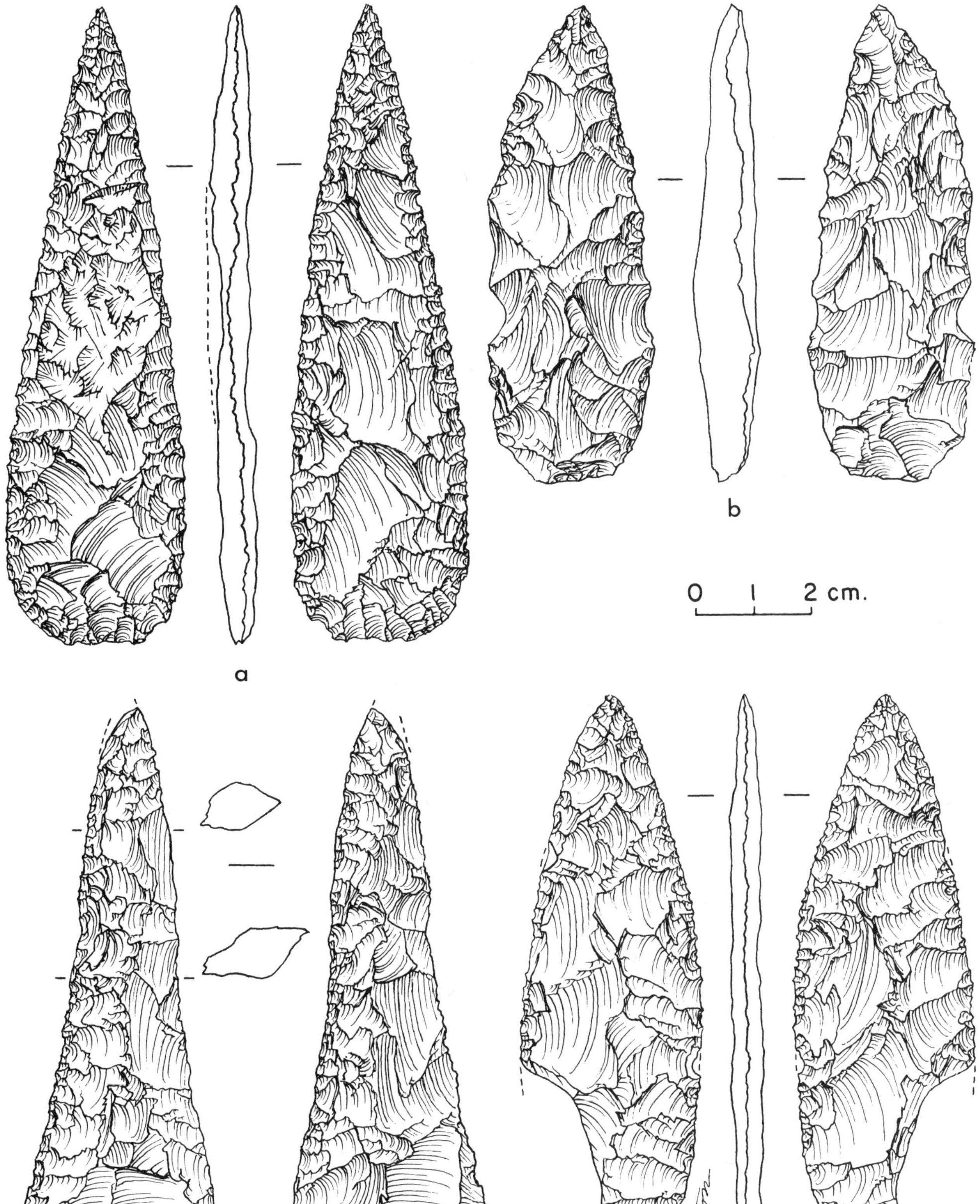
a
b
0 1 2 cm.
c
d

Figure 45. Trianguloid bifaces and bifacial knife. *a–d*, *f*: trianguloid bifaces; *e*: beveled, bifacially worked, elongated knife. *b* and *e* needed both transverse sections and edge-only profiles to show shift of thickness and edge slope, details not possible to give adequately by merely front and back views.

Sources: *a–d*, *f*: Henry et al. 1980; *e*: Wendorf and Schild 1980.

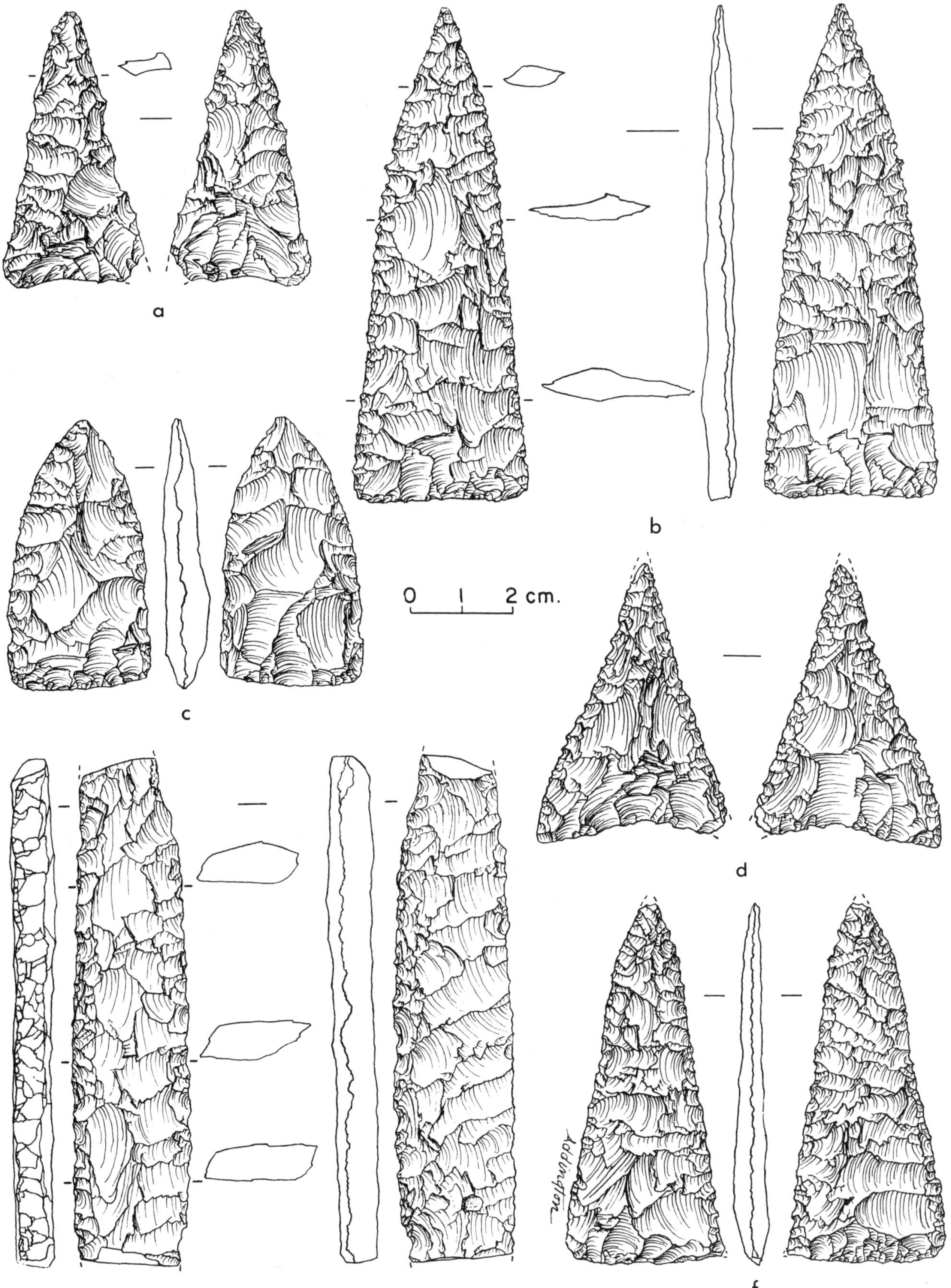
a
b
c
0 1 2 cm.
d
e
f

Figure 46. Drills and additional bifaces. *a*, *f*: drills; *b*: lanceolate biface; *c*: small ovate biface; *d*, *h*: amygdaloid bifaces; *e*: ovate discoidal biface; *g*: drill with modified base; *d*–*f*: show traces of older, patinated flake scars.

Sources: *a*, *c*: Henry 1978a; *b*, *d*–*h*: Henry et al. 1980.

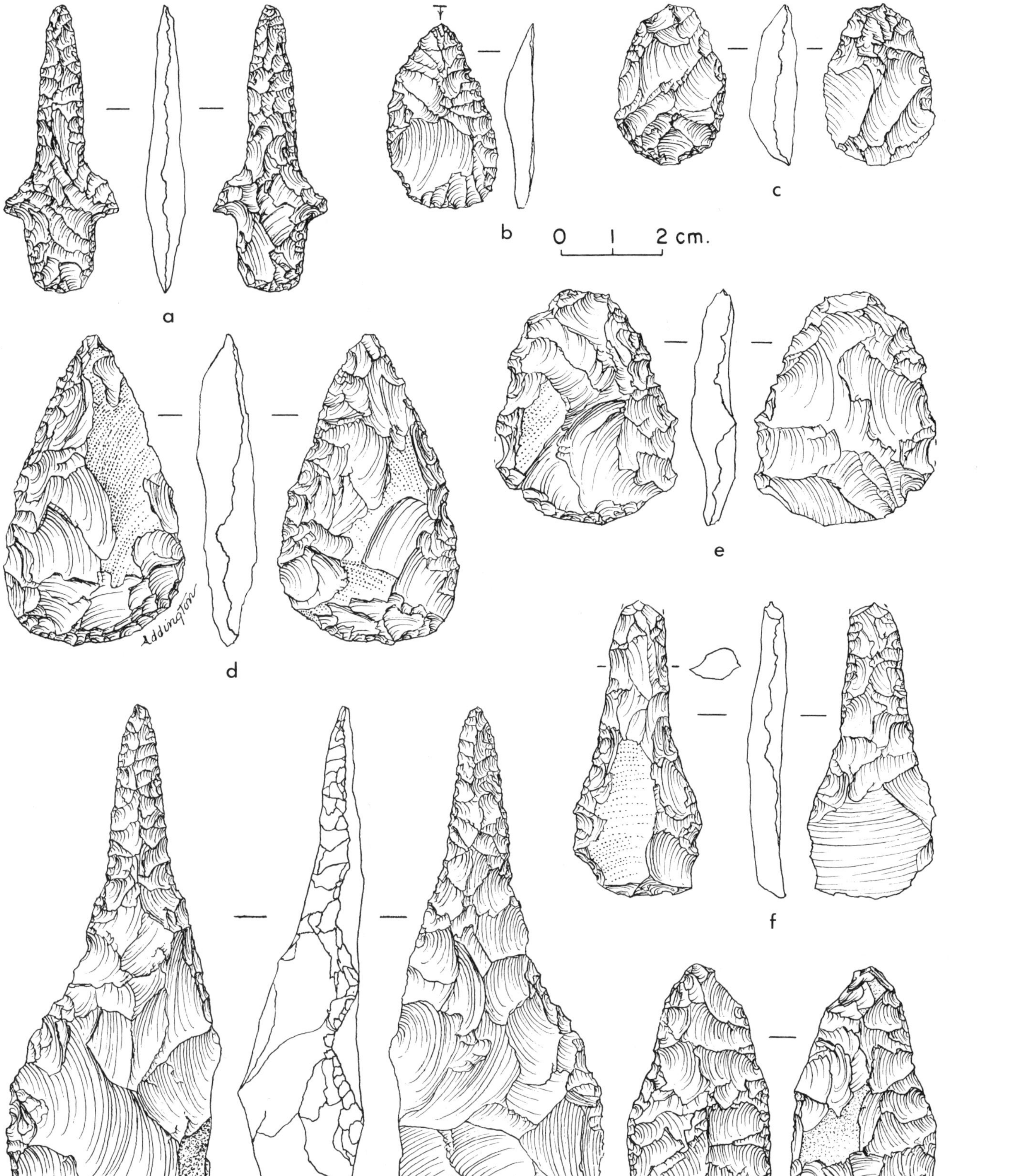

a
b
c
0 1 2 cm.
d
e
f
g
h
Addington

Figure 47. Bifaces. *a*: amygdaloid biface; *b*: biface. These were drawn to withstand even more drastic reduction than used here.

Sources: *b*: Henry 1982; *a*: Schild and Wendorf, eds. 1977.

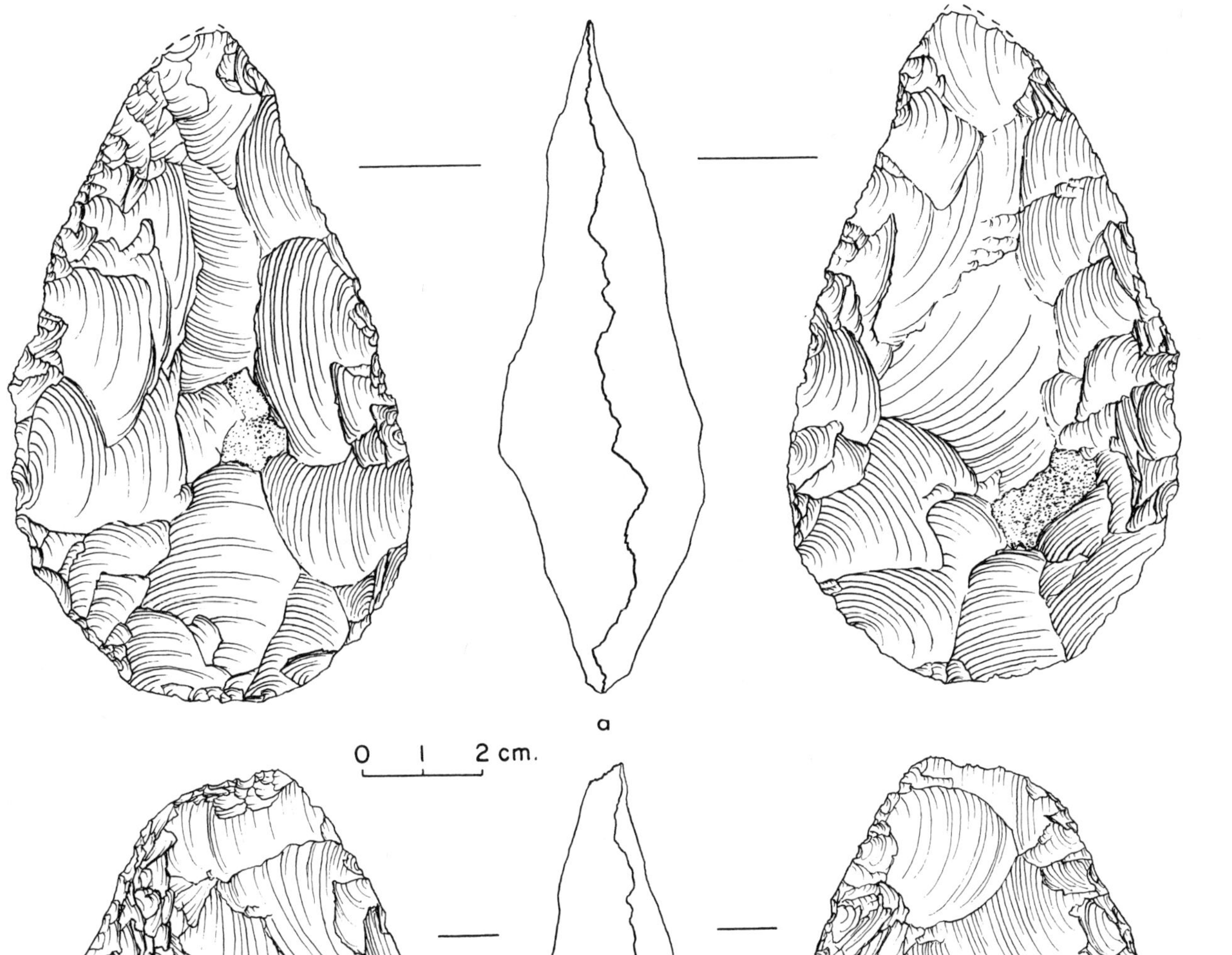
a
0 1 2 cm.
b
Addington

Figure 48. Amygdaloid biface. Originally published at approximately two-thirds this size, it is reproduced here at natural size to demonstrate "defensive" drawing for anticipated drastic reduction.

Source: Schild and Wendorf, eds. 1977.

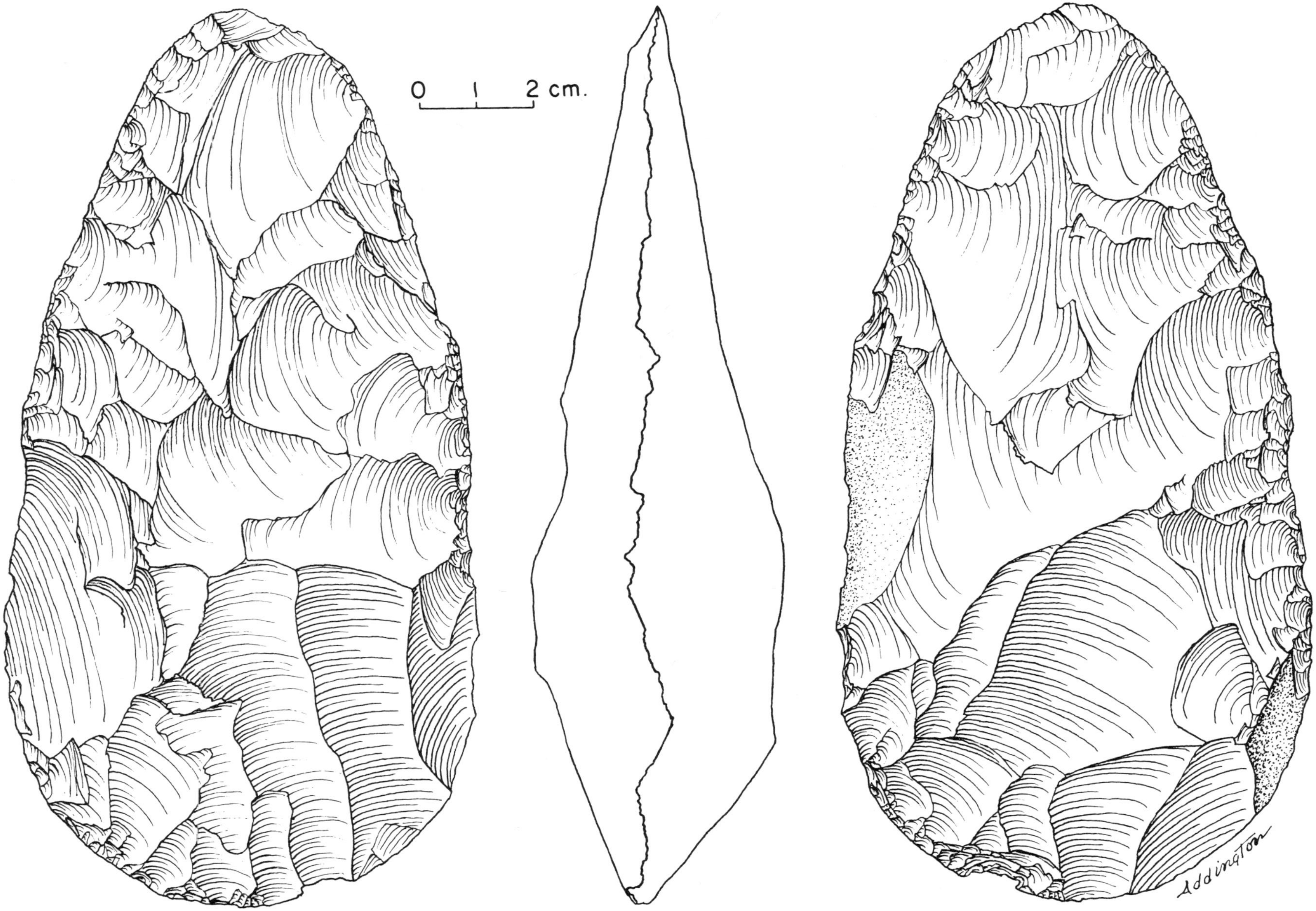
0 1 2 cm.
Addington

Figure 49. Bifacial choppers.

Source: Marks, ed. 1976.

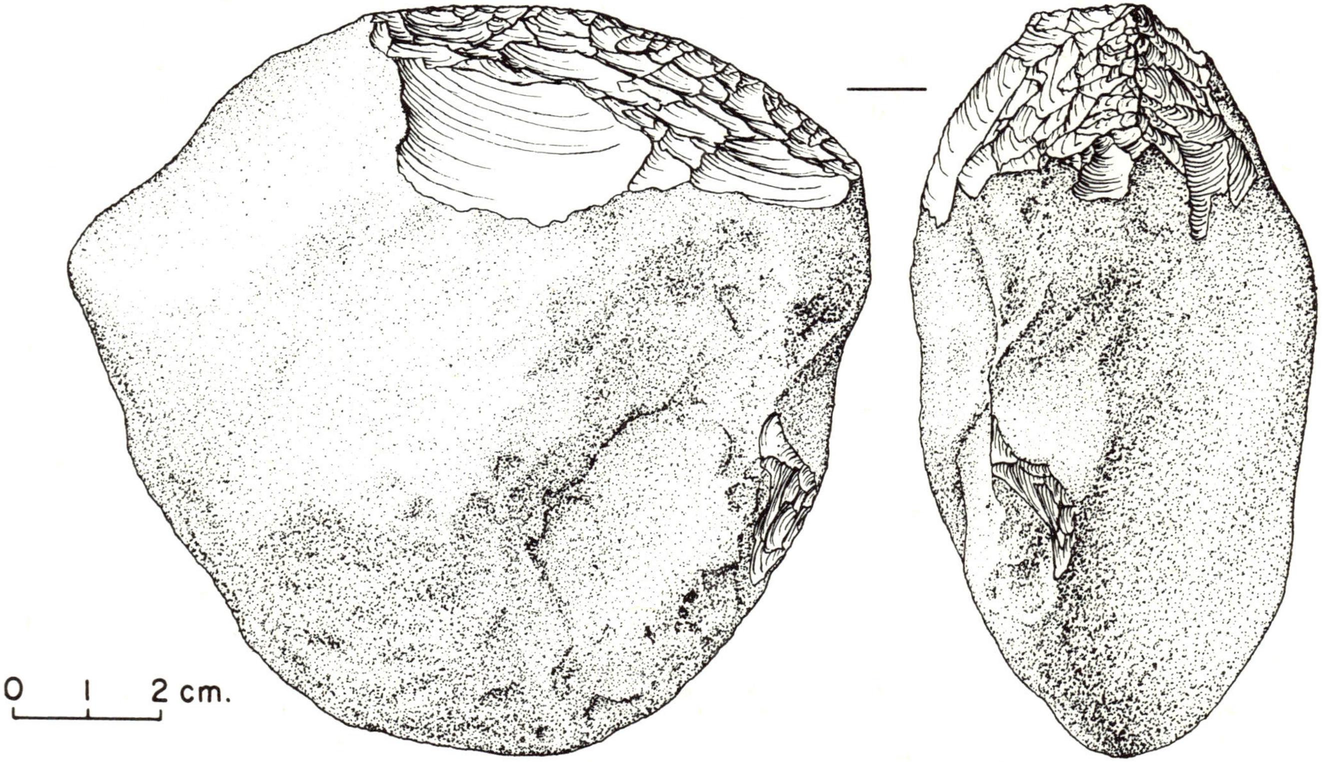

a

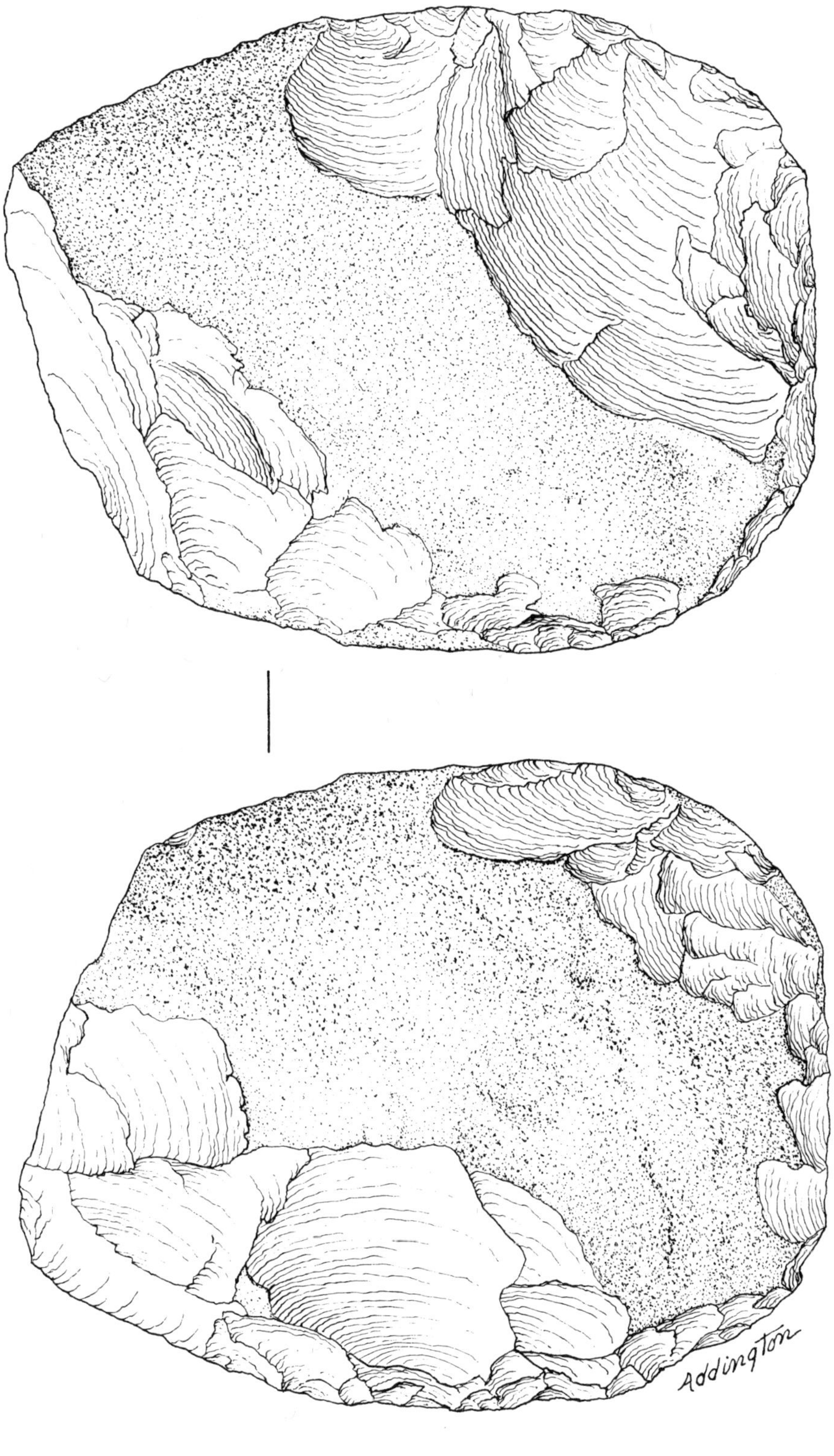

b

Figure 50. Unifacial chopper.

Source: Marks, ed. 1976.

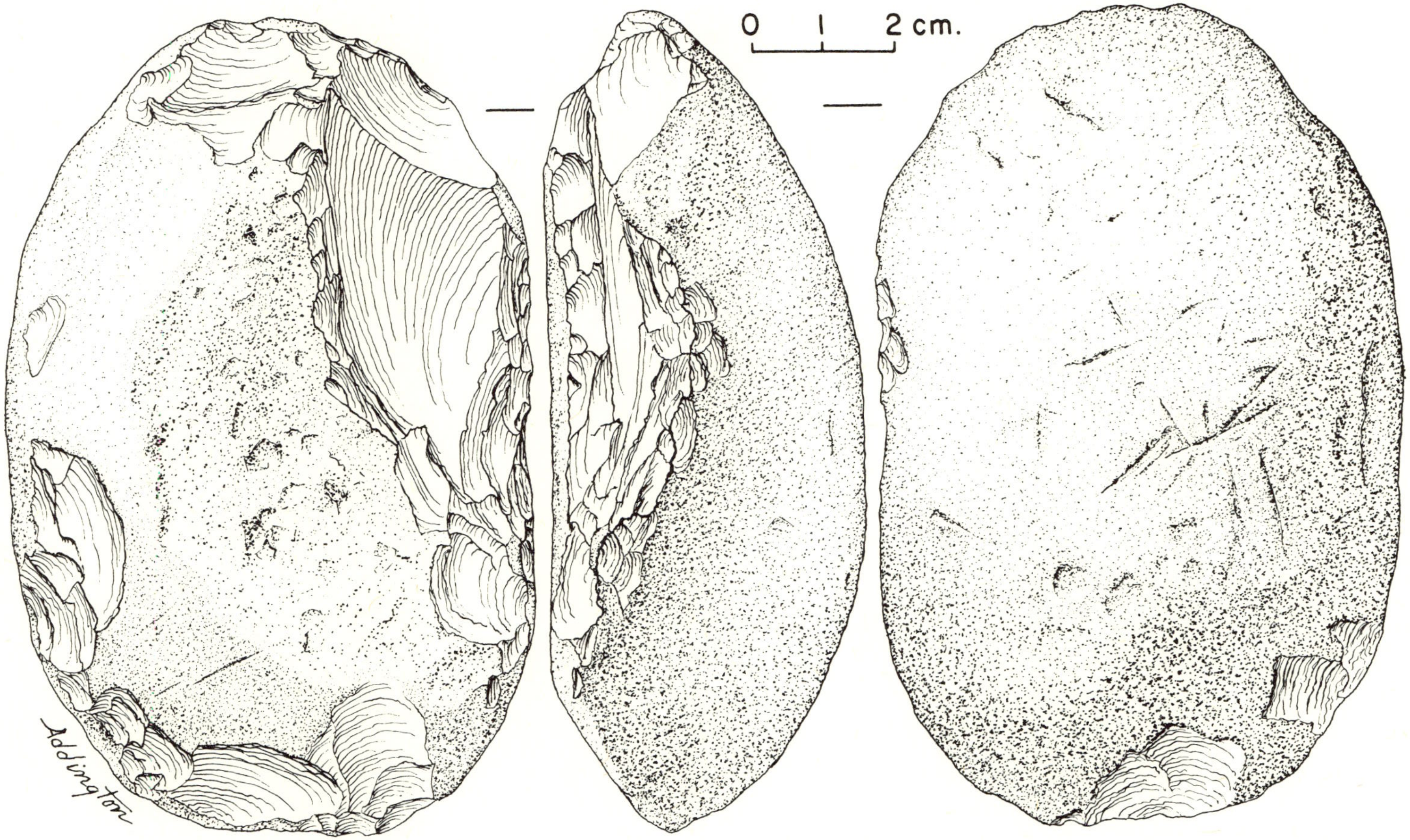
0 1 2 cm.
Addington

Figure 51. Hammerstones. *a*: hammerstone on a core; *b*: chopper/hammerstone on a cobble.

Sources: *a*: Marks, ed. 1976; *b*: Marks, ed. 1977.

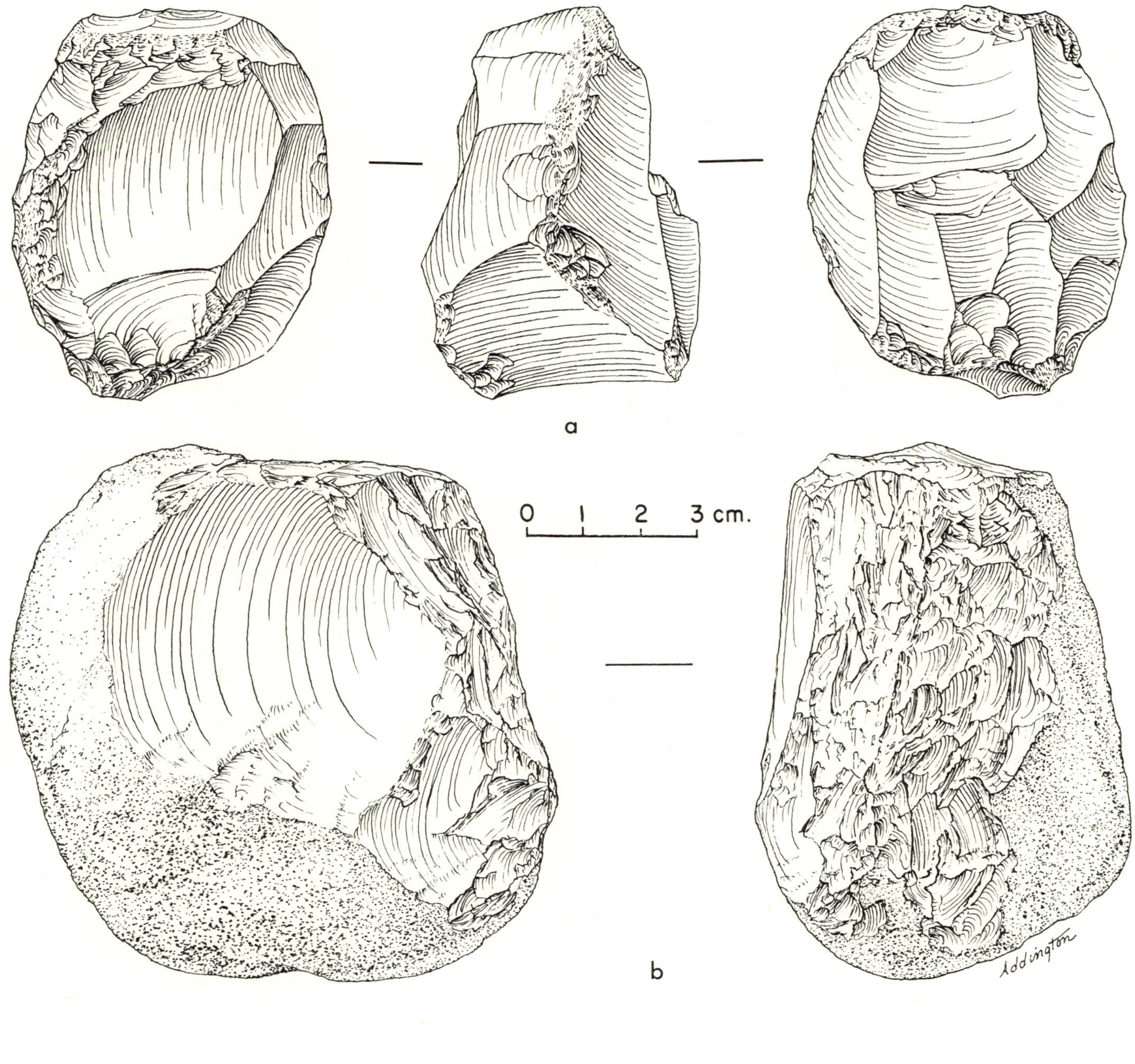
a
0 1 2 3 cm.
b
Addington

Figure 52. Battered pieces. *a*: on a large primary flake with a distorted ventral surface visible in profile view; *b*: inversely battered; *c*, *d*: bifacially battered. Note; if the partial view of the ventral surface of *d* had been drawn tucked behind the right lateral edge of the major view, the layout flexibility of the drawing would have improved upon the awkward effect above.

Sources: *b*, *d*: Marks, ed. 1976; *c*: Marks, ed. 1977; *a*: Marks, ed. 1983b.

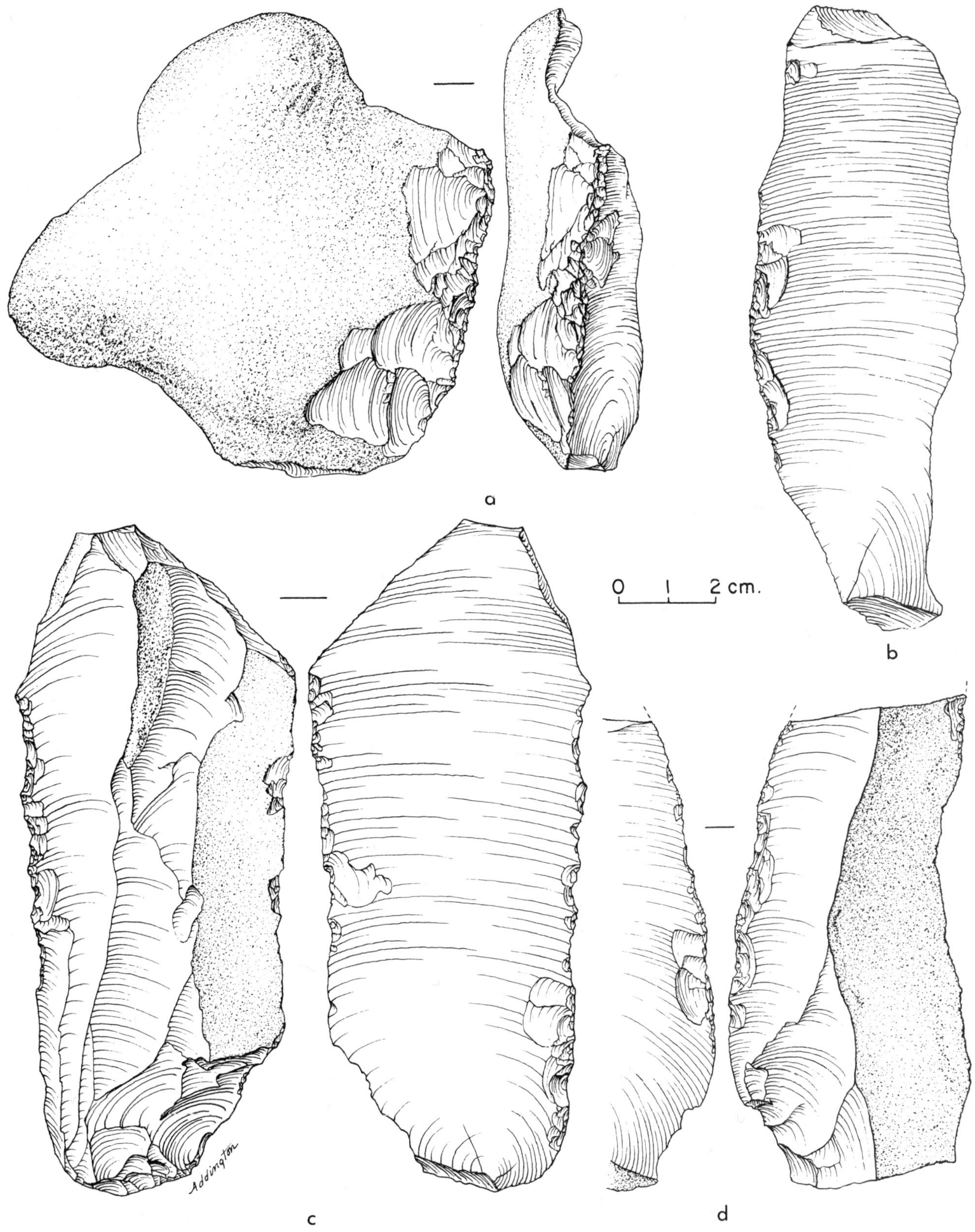
a
b
0 1 2 cm.
Addington
c
d

Figure 53. Sickle sheen pieces. *a–d*: to save time and space, sickle sheen occurrences have been stippled in areas already containing direction lines; *e–i*: repeat outlines contain the stippling of the sickle sheen, a method preferred for its added clarity.

Sources: *a–g*, *i*: Marks, ed. 1977; *h*: Wendorf and Schild, assemblers, and Close, ed. 1984.

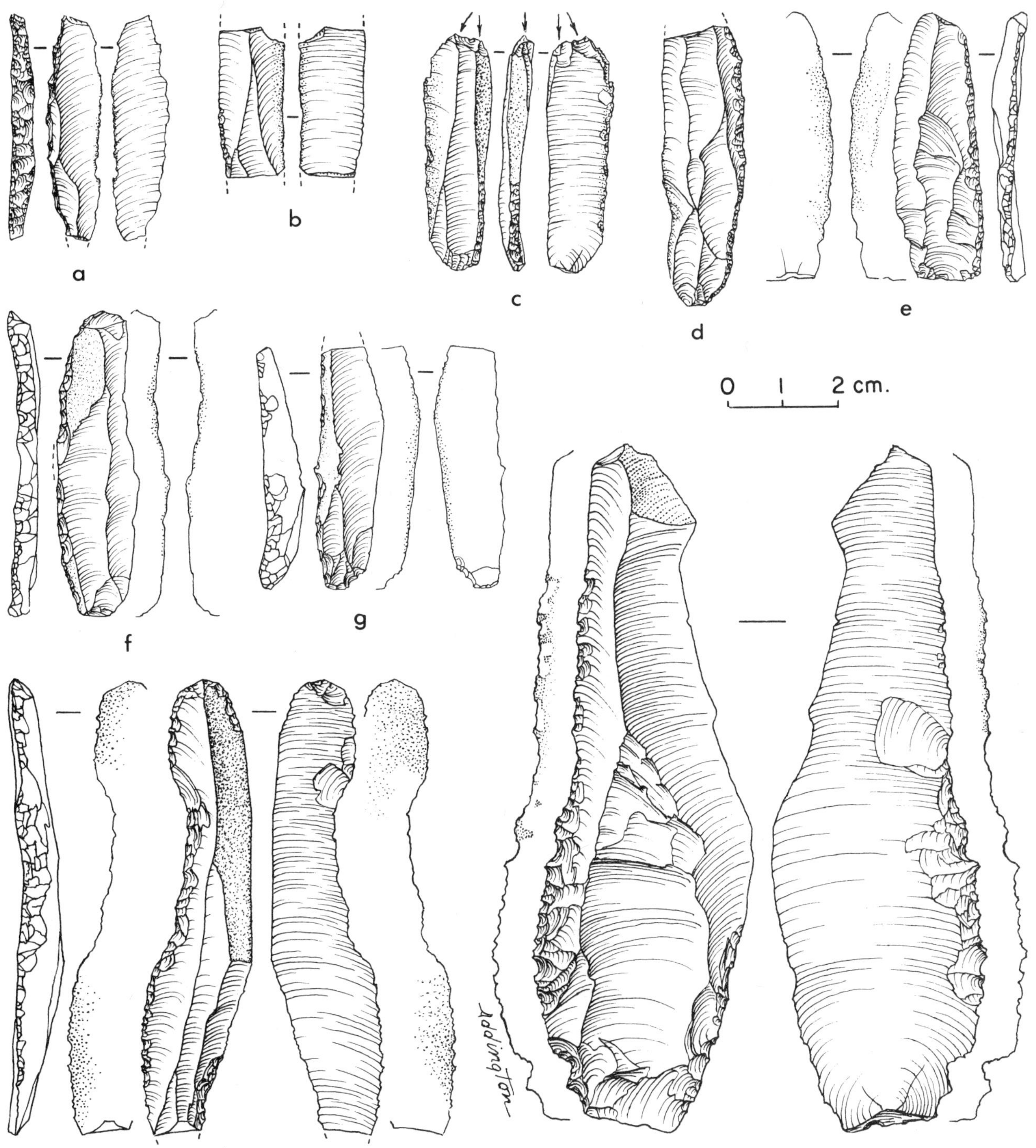
a
b
c
d
e
f
g
h
i
0 1 2 cm.
Addington

Figure 54. Ground stone. *b*, *d*, *e*: celts; *a*: sandstone smoother with scratches; *c*: handstone with ground surfaces. *b*: celt with basal break retouched; *d* and *e* show some pecking as well as grinding. In the profiles of *b*, *c*, and *e*, note how the light and dark distribution of the stippling can give the effect of a sudden or a gradual transition from the front surface to the back surface. A nearly three-dimensional effect is achieved by keeping the darkest stippling at the area of change from "front" to "back."

Sources: *a*, *c*: Marks, ed. 1977; *b*, *d*, *e*: Wendorf and Schild 1980.

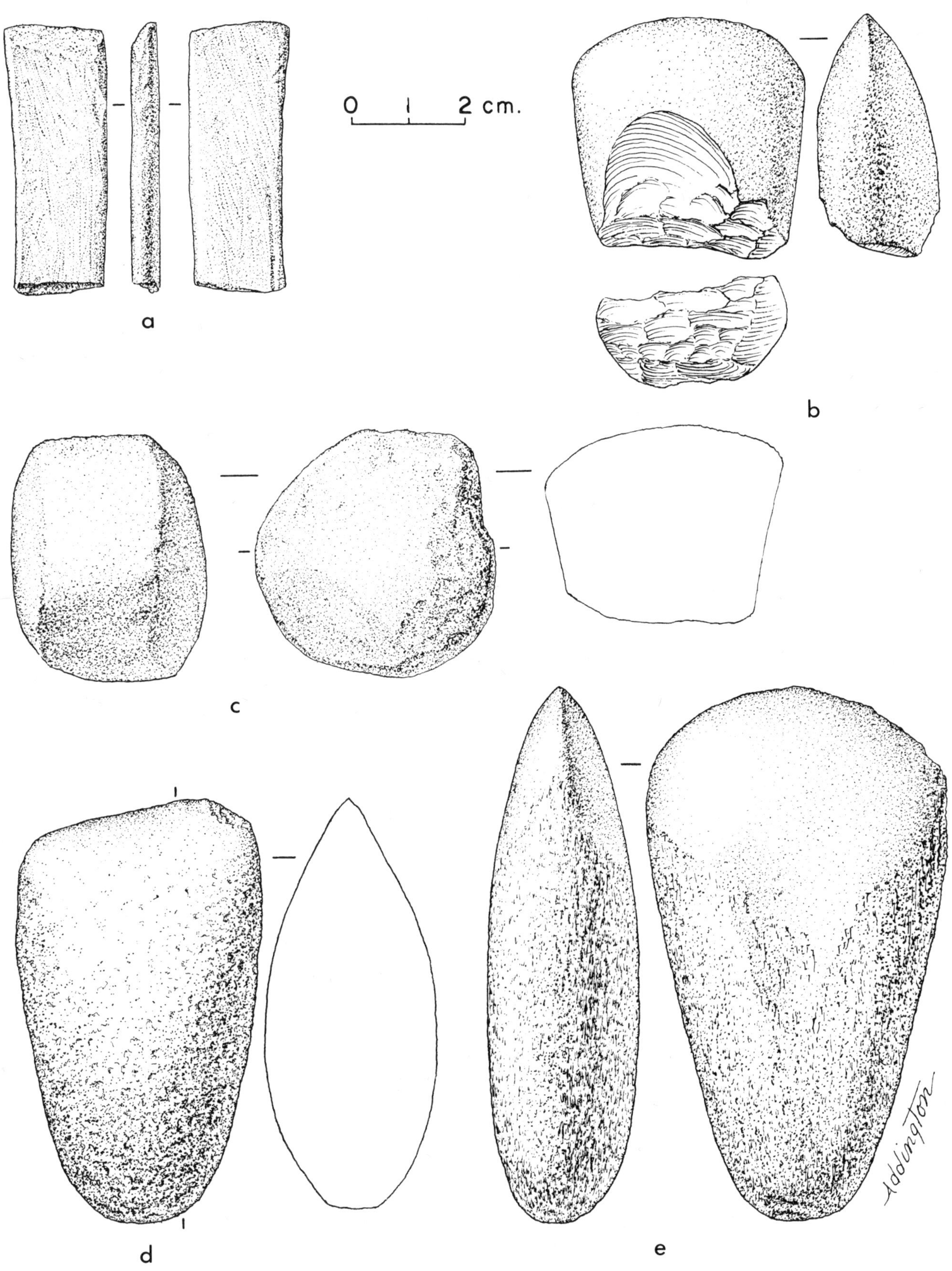
0 1 2 cm.
a
b
c
d
e
Addington

Figure 55. Grooved stone and ground stone. *a*: grooved stone with nonparallel striations; *b*: ground stone with pecking and pits.

Source: *a*, *b*: Marks, ed. 1976.

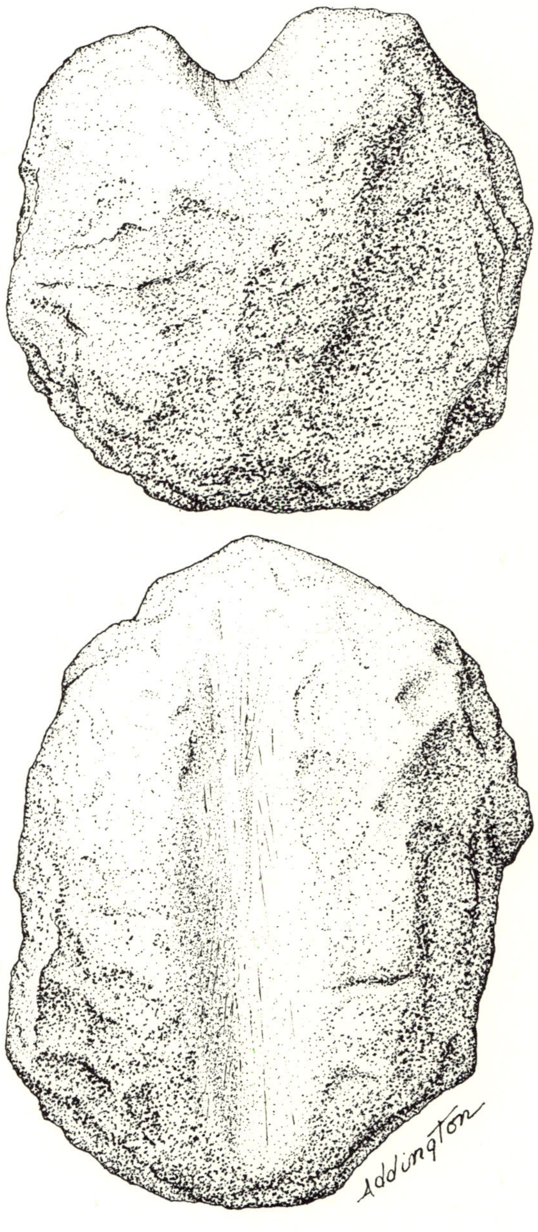

a

b

Figure 56. Layout of assorted core types, shown here at natural size. *a*: opposed platform core; *b*: pyramidal core; *c*, *d*, *g*, *h*: single platform cores (*h* has opposed preparation scars and *d* has a cortex platform); *e*: change of orientation core; *f*: discoidal core. Composition notes: *a* is "strong" at the bottom and therefore more suitable above the "weak" slopes of the top of *c*; conversely, *d* is "strong" at its top and therefore helps the "weak" bottom of *b*. Though *e* is rather small, its three views extend far enough to break what could have been a white "river" if a "two" had monotonously stacked "twos" down the left. And though *f* is more massive than *h*, the triple reiteration of the three views of *h* makes it weightier and better suited to the bottom row in this case.

Sources: *c*, *f*, *g*, *h*: Marks, ed. 1976; *a*, *b*, *d*, *e*: Marks, ed. 1977.

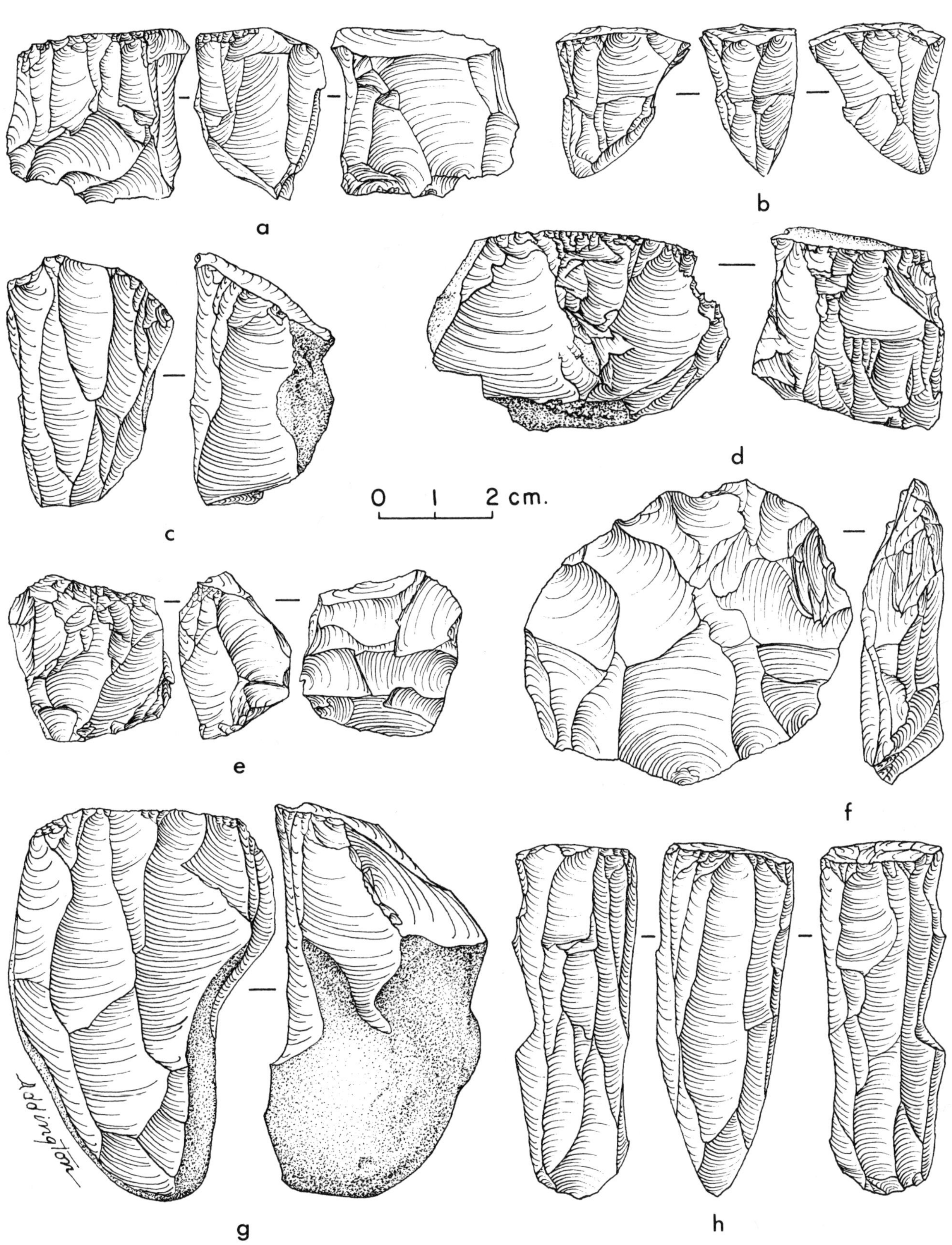
a
b
c
d
0 1 2 cm.
e
f
g
h

Figure 57. Layout of assorted core types, here reduced approximately 15 percent. *a*: single platform core that needed the third view to show multidirectional scars of the platform; *b*: opposed platform bladelet core; *c*: single platform blade core whose scar remnants in the right view show that platform rejuvenation eliminated their points of impact before the major removals in the left view were effected; *d*: opposed platform, opposite side, blade core, with thermal damage pits in the cortex; *e*: opposed platform blade core; *f*: change of orientation core; *g*: opposed platform core, with some older patinated scars; *h*: opposed platform blade core, with much preparatory flaking (left view) to shape the front face of the core for removal of blades. Composition notes: the bottom three rows are examples of "twos" and "threes" filling rows; different heights of *e*–*h* are arranged to interlock, but the nearly-even bottoms of row two and tops of row three threaten to snap the layout apart horizontally between them, blocked somewhat by placing the scale in that area.

Sources: *a*–*f*: Marks, ed. 1976; *g*–*h*: Marks, ed. 1977.

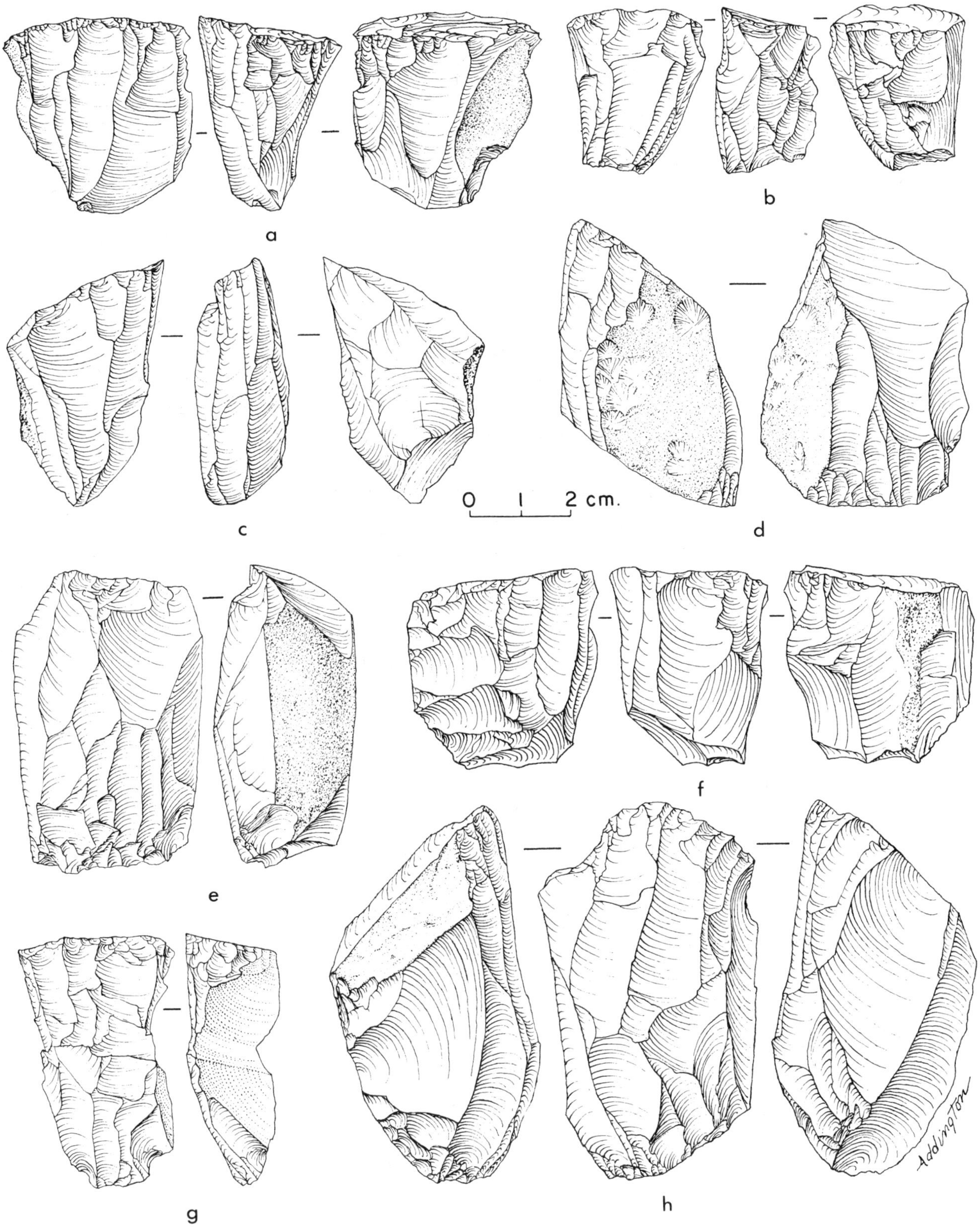
a
b
c
0 1 2 cm.
d
e
f
g
h
Addington

Figure 58. Levallois and Halfa cores. *a*, *c*: Levallois; *b*, *d*: Halfa. By convention, these two types are always oriented with the major scar's point of impact at the bottom of the drawing (meaning oriented so that the force of detachment that formed the major scar appears to have been directed upward). Composition notes: *c* and *d* are examples of multiple views of large pieces drawn to fit their original layout's width; *a* and *b* are examples of "threes" alternating to touch side margins when their multiple views form units too wide for the same row and there are no other pieces to fill the width—not very satisfactory when the views have been drawn as close together as these, but often better than centering each "three" in a pyramid stack. The scale offers feeble help.

Sources: *c*: Marks, ed. 1976; *a*, *b*, *d*: Wendorf and Schild 1976.

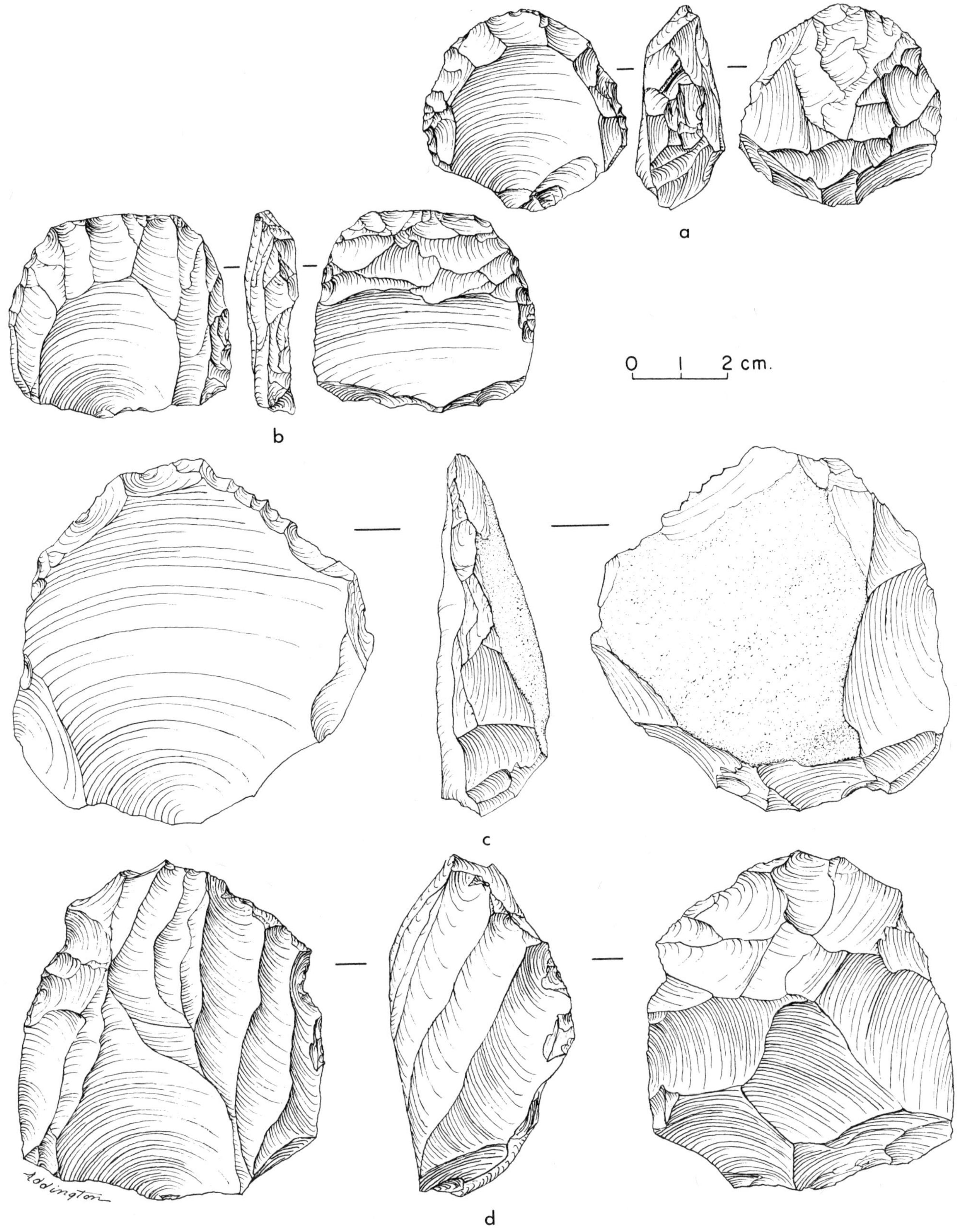
a
b
0 1 2 cm.
c
d
Addington

Figure 59. Cores, here natural size, with their multiple views drawn and spaced to fit exactly the layout width of their original publication. *a*: pyramidal core; *b*: opposed platform blade core.

Sources: *b*: Marks, ed. 1976; *a*: Marks, ed. 1977.

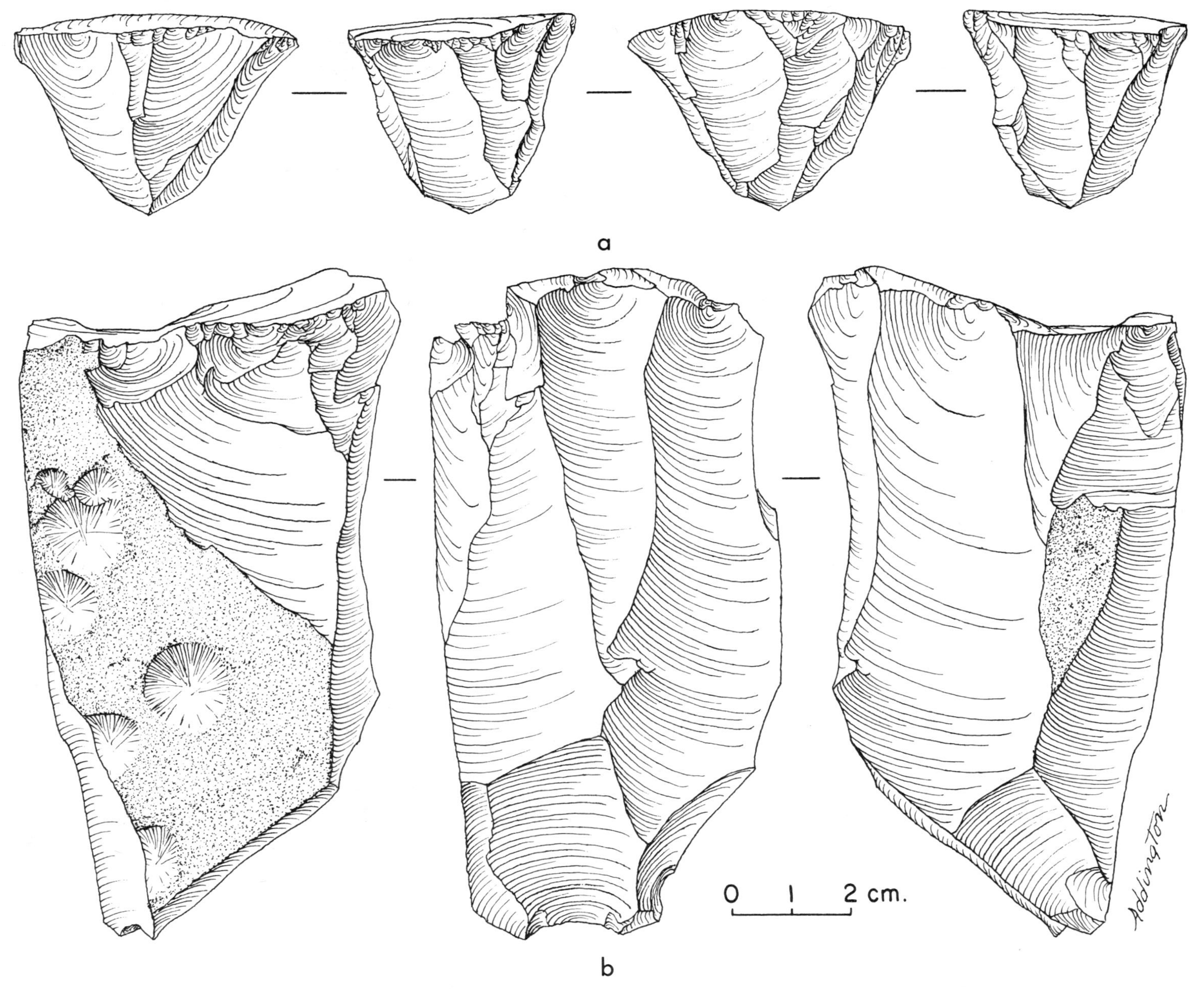
a
b
0 1 2 cm.
Addington

Figure 60. Reconstructed core. The heavy lines show where the separate pieces fit together and have been glued.

Source: Marks, ed. 1983b.

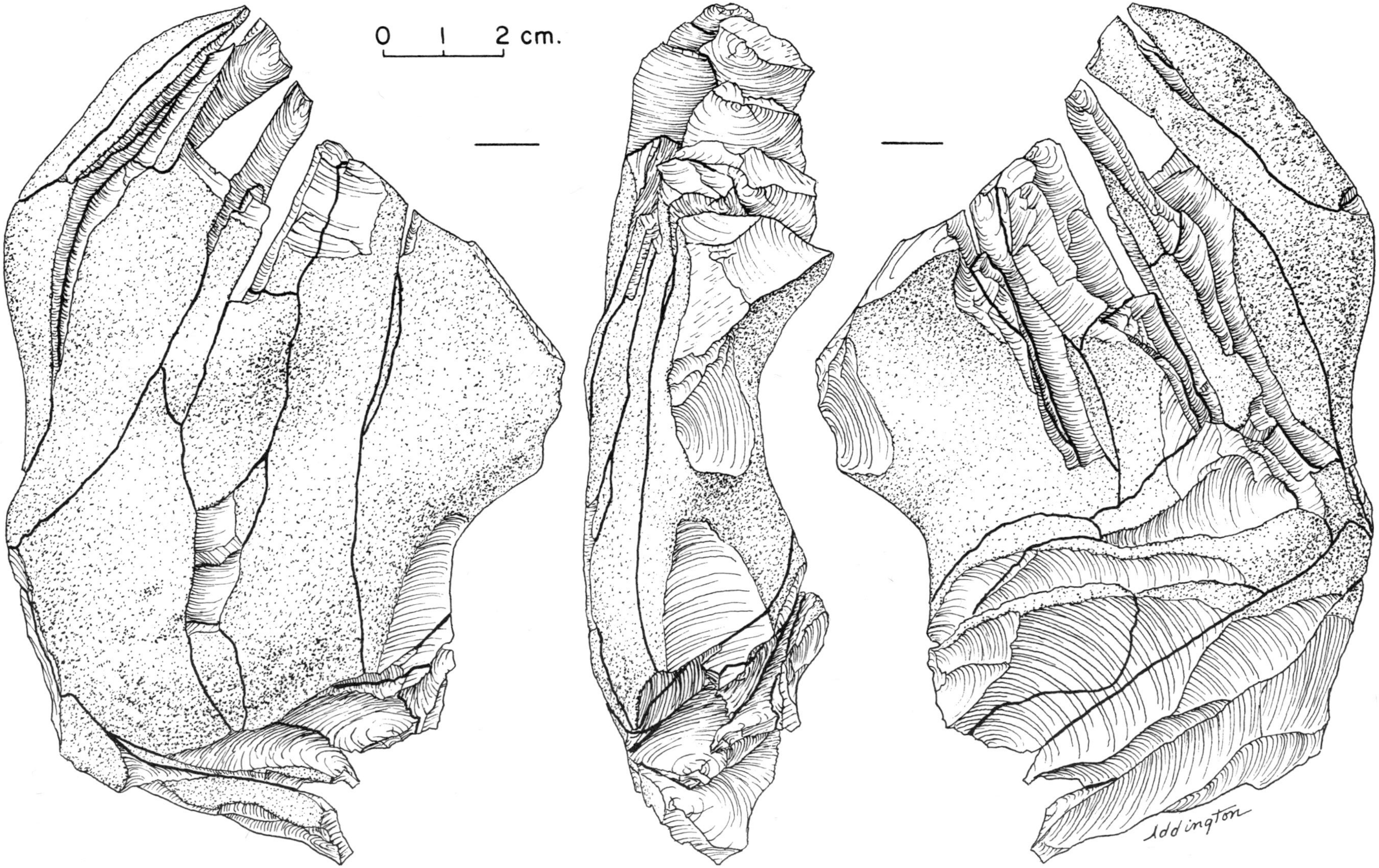
0 1 2 cm.
Addington

Figure 61. Front, back, and top views of reconstructed core.

Source: Marks, ed. 1983b.

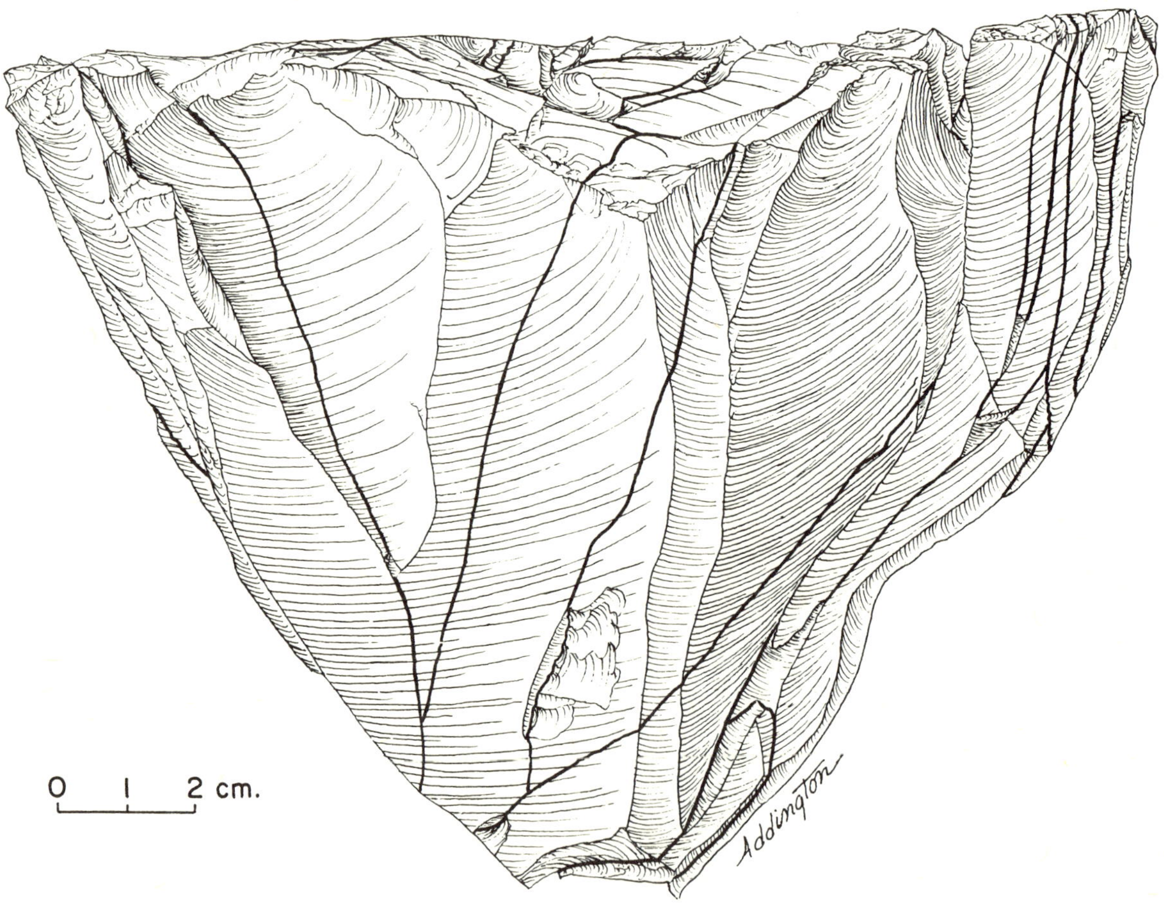

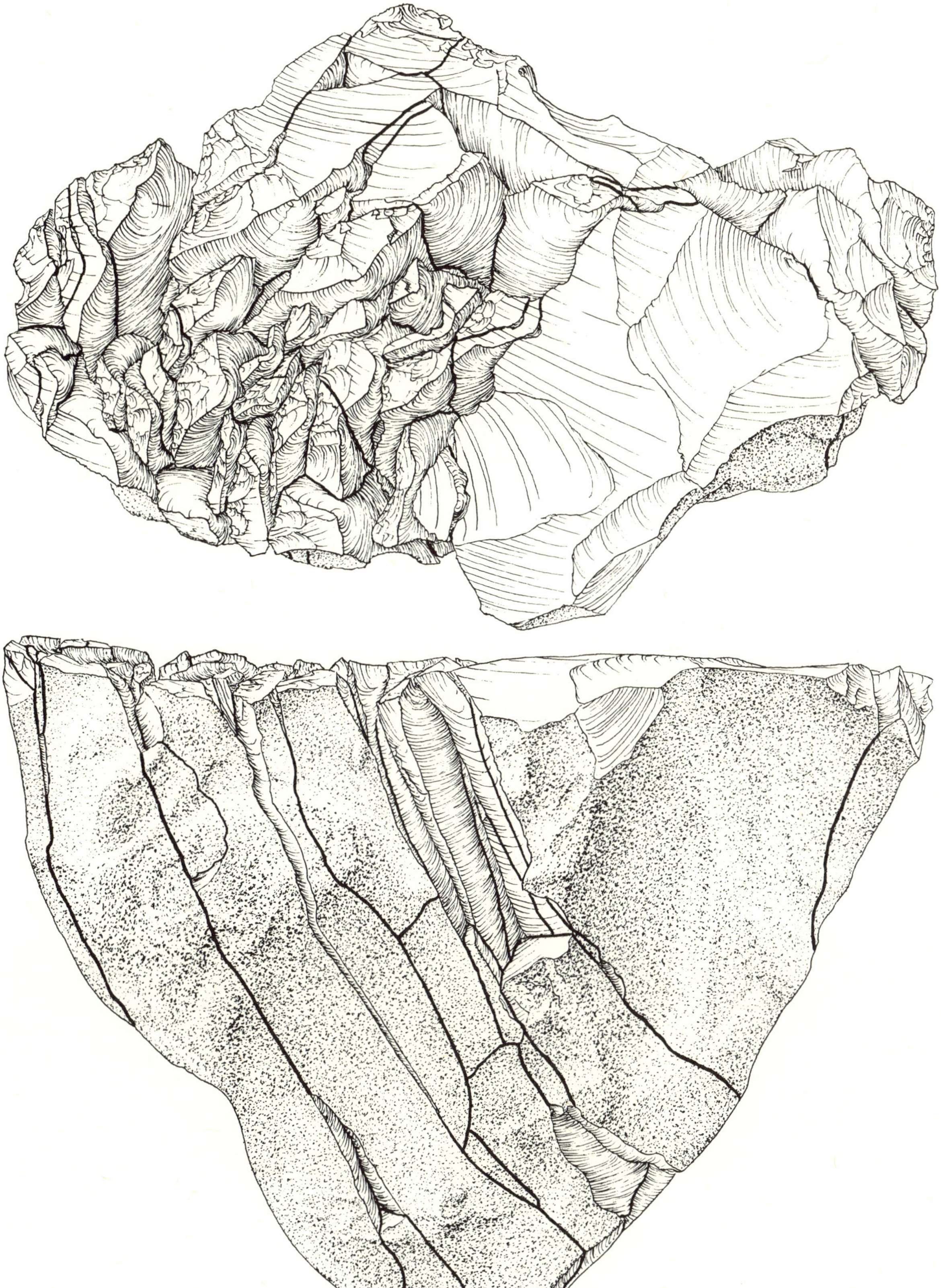

Part 2

The Drawing Process

4 The Pencil Stage of the Drawing

All the foregoing material--on terminology, the codes of conventional orientations, views, symbols, the marks to use, artifact examination--has supplied you with the preliminary background information for lithic illustration and has brought you to the drawing process itself. Read through this entire chapter before putting pencil to paper. A maddening peculiarity of this kind of illustrating, and a characteristic that complicates the teaching of it, has become apparent by now: you must know about everything before you can do anything.

Draw artifacts in pencil first, for the obvious reason that corrections are to be expected. Unless the illustrator is experienced in typology and in "reading" rock surfaces, the pencil stage of the work should be checked for accuracy by the archeologist. After a few days of penciling, have the accumulated drawings approved; then ink them before intimate acquaintance with the artifacts fades. When to stop penciling and start inking is flexible, influenced by such factors as the overall number of pieces to be illustrated, their size and complexity, the experience of the artist, and the availability of the archeologist to approve the pencil stage of the drawings. If the job calls for dummy layouts, pencil all artifacts in the collection with a softer leaded pencil, a 2H that will pick up on an ordinary photocopy, and do not ink any of them until after the dummy layouts are made. Detailed instructions are given in the chapters dealing with layouts.

Have the proper lamplight placed so an artifact held under it will be illuminated correctly from the upper left. A magnifying glass, adjustable in angle and attached to a weighted base, is a necessity. It can be rigged from an ordinary reading glass of the "Sherlock Holmes" type if a ready-made one is unobtainable.

Keep your pencil well sharpened for the precision required by small flake scars, but restrain the tendency to become so emphatic with minor corrections that pencil grooves will interfere with the pen; better to erase a gently made error using an erasure template or shield that exposes only the bit of line to be erased and prevents blurring the rest. Swipes with the edge of the hand are not the best technique to brush away eraser crumbs; a short-handled drafting brush does a better job without smearing fine detail. This is a practical caution; the time spent marking, reading, and drawing with precision, say, an area of complex retouch scars will make the inking process go faster, but not if the pencil lines become so blurred that their identities must be puzzled out again.

Make a habit of rotating the piece of drawing paper to any different position that affords the most comfortable angle for drawing a particular set of lines or allows a better approach to an edge for drawing its outline. However, with all the turning, remain aware of which is the bottom of the drawing.

When the drawing has progressed to the point where the artifact must be held in the light at the correct angle so the location of shadows on its surface can be noted, use a couple of strips of drafting tape on opposite edges of the paper to anchor it to the table and free one hand to hold the artifact. Drafting tape sticks just enough to keep the paper still, but it is easy to pull loose the paper and tape together, turn the drawing to a different angle, and press the tape to grip the table again.

If a line is meant to be continuous, draw it so; a sketchy, "hairy" pencil line in this kind of drawing is imprecision that will cause you trouble later. Your drawing must be an explicit statement of data, not an impressionistic sketch.

When a piece is so bowed, twisted or unbalanced that it tends to tilt over to one side instead of presenting a true "front view" as it lies on the paper, prop it in position for the first steps of the drawing process. Either pieces sliced from an eraser or, more satisfactory, pellets of dried rubber cement can be poked under the edge of the artifact, out of the way of any retouch that must be drawn at the outline. It is a convenience to keep on hand a collection of assorted sizes of these props, from pinhead to walnut size. The balls of rubber cement stay in place better than eraser bits. With the brush attached to the bottle cap, paint some rubber cement on a sheet of paper, let it dry a few minutes, then rub with a finger to form pellets. A word of caution: do not use modeling clay to prop artifacts. Its oil clings to them and ruins the drawing paper. No holder that immobilizes the artifact is suitable; the piece must be handled and turned repeatedly under the light to examine the flake scars while drawing is in progress.

Learning to look with one eye may require practice. To see what problems you will face if you do not take the word of only one eye as you draw, make this test. Hold an eraser or other small object about an inch above a printed page; peering straight down on the object and page, without moving your head, look at the print with first one eye, then the other. The print and the object will appear to change position drastically in relation to each other. This stereoscopic fact of life can give a beginner trouble in translating a three-dimensional object into lines on a piece of paper.

Following is the sequence for penciling an artifact. Procedures that are merely mentioned in the list are fully explained elsewhere in this manual.

1. Examine the piece, decide the orientation, and select the kinds and number of views needed.

2. Cut a piece of drawing paper large enough to hold all the anticipated views of a single artifact, plus at least an inch of margin all around.

3. With the side of the pencil lead, carefully blacken the flake scar ridges that will be visible in the first and major view. Done with precision, this process makes the exact outline of every scar equally visible at one time and from one position. Indeed, small continuous retouch scars along an edge can be impossible to distinguish, when the piece is in position on the paper, unless their ridges are marked with pencil lead. This technique eliminates so much of the difficulty of drawing extensive retouch, and thick pieces whose foreshortened scars are a problem, that it should be routinely followed for all pieces. On artifacts made of black or dark brown material, use a colored pencil, such as light green or orange, to mark the scar ridges.

4. Position the artifact on the paper according to the correct orientation for its tool type and the dictates of its individual morphology and pencil its outline, making an effort to copy the characteristics of its edge. The "quick" way of holding the piece down with one hand and dragging the pencil around it does not produce an accurate outline.

5. Leave the artifact in position, look straight down on it and select a surface feature in the central part, such as where a hinged-off scar meets another scar ridge. Visualize an imaginary line from the feature straight out to the side, beyond the outline, and lightly pencil that feature. It might, in this example, look like a letter T. Maintaining exactly the same point of view directly over the piece, figure another imaginary line from the same feature straight up or down and, beyond the outline, pencil the same small T shape. Go through this routine with a few additional scattered features, selecting ones whose accurate placement will help in drawing the areas around them. When several such landmarks have been located by this method, backtracking the pairs of imaginary lines will place the features at the correct points within the outline. This system of locating key points in the complex pattern of flake scars eliminates much trial and error. For now, leave the artifact in place and let the backtracking wait until a later step in this sequence.

6. With the piece still in position on the paper, move one edge of the artifact about one-eighth inch back from, its outline, but still exactly parallel to it, and copy whatever retouch or scar ridges join the outline. If the retouch is more invasive, move the piece again a little farther from the outline of that edge and continue copying whatever additional retouch scars are revealed. Go through this routine with each edge of the piece. Now you have drawn all the smallest scars and portions of scars that are in the vicinity of the outline of the piece.

7. Remove the artifact from the paper and finish drawing any incomplete small scar outlines. Now, backtrack the imaginary vertical and horizontal from each pair of the little feature markers you penciled outside the outline to the spot where they meet, and there draw the feature that the pair represents. With those strategic bits in place, the remaining flake scar outlines are less trouble to complete. Erase the little markers outside the drawing; they have served their purpose.

8. Pencil the "direction lines" into the flake scars by following the instructions under that listing in chapter 1. If cortex is present, indicate it with some loose scribble as explained there under "stippling." Treatments of other less common surfaces can be found in the same section.

9. Most pieces need multiple views to present the data adequately. Always rule parallel horizontal guide lines touching the bottom and top of the first-drawn (usually dorsal) view. (Even if no more views are needed, the lines will aid layout makeup later.) If profile and/or ventral views are desired, add the rest of the horizontal guide lines as described in chapter 2. Then draw the necessary additional views to fit between the top and bottom guide lines. In drawing a profile, note that wherever the back of the piece touches the drawing paper in the "front" view, those places are drawn to touch the vertical guide line against which the profile drawing is set. Also, remember that a profile is drawn beside the edge whose data it amplifies. Space the views about one-eighth inch apart for very small pieces, and about one-fourth inch apart for medium and large ones. Too widely spaced views of the same piece lose their visual relation to each other and result in poor layouts. An exception to that spacing: large, thick pieces whose multiple views could be made to fit exactly across a known layout width by slight increase or decrease of the space between views. This little forethought in drawing will save time later when the illustration is found to fit neatly into the layout without adjustment.

10. Pencil the complete site number, artifact identification, and tool type on every drawing, without exception. Make this an invariable rule. It is important to leave about one-fourth inch between the bottom of the drawing, whether base line or butt view is the "bottom," and the identification data. The reason for this has to do with the making up of layouts, discussed in later chapters.

5 Inking the Pencil Drawing

The finished pencil drawing, in summary, has all the flake scar shapes in place; curved direction lines in the scars; cortex, if present, designated by some loose scribble; and any other special surface treatments indicated. The areas of shadow on the piece have been suggested by the spacing of the penciled direction lines but not drawn in full intensity of contrast. Symbols, if necessary, have been supplied. Hyphens, have been ruled between multiple views that are in horizontal series. All the guide lines are still in place and should remain until their erasure is specified at a later point. The complete identification of the artifact is penciled below the drawing, at least one-fourth inch away.

If this pencil drawing is part of a project that requires dummy layouts, do not ink it yet. Instead, follow the instructions in chapter 7 detailing a different schedule of procedures for dummy layouts.

If, on the other hand, this drawing is a part of the more common pencil-ink-layout type of project, inking the pencil drawing is the next major step. First, however, a description of the recommended drawing instrument and some guidance in its use will be helpful.

The small, simple steel penpoint--the one with which all the illustrations in this manual were drawn--is a Hunt's no. 104 that slips into a penstaff of the same brand and number. Superficially it resembles the more widely known "crow quill" penpoint, but it has an important difference: the shorter tip of the 104 makes it more rigid, a characteristic that gives the illustrator greater control than does the sensitive flex of the crow quill. Nevertheless, the 104 is capable of a degree of variation in the line it produces.

The free ends of lines will taper if they are drawn with some "follow through" in a rather brisk movement that might be described as a controlled "whisk," as opposed to a cautious creep along a pencil line to a dead stop at the end. Tapering the many free ends of direction lines contributes to smooth gradation of shadows. Important to the appearance of a finished drawing, the repeated arcs of the direction lines will blend and flow in an effect it is not possible to produce successfully with the reservoir type of drafting pen, which is designed to execute a line of uniform thickness. Although reservoir pens may seem less trouble to manipulate than the simple steel penpoint--whose design is only a step away from the well-trimmed quill of a goose feather--the more technologically advanced instruments are not suitable for this kind of drawing. Most artifact illustrations drawn with drafting pens are mechanical looking and graceless, without individual emphasis or character. They have the faceless anonymity of diagrams. Use the little 104. Practice with it on a piece of the same drawing paper to become acquainted with how it feels marking on that surface; discover what it will and will not do.

The point cannot be *pushed* in an arc; it will hang in the paper. It must be *pulled* somewhat toward you (a "sagging" curve, in the case of direction lines). That is the reason for turning the paper as often as needed during the inking process to afford easy angles for placing the arcs. Practice a continuous nested series of direction lines, varying their lengths as you go, "pulling" a long series from the left, and then another series from the right. A few minutes of alternating this way will begin to educate your hand in a new set of movements and tell you if you need more practice with the pen before you start to ink the pencil drawing.

You can increase your control of the pen if you will keep the entire forearm on the table, in something of a relaxed sprawl, and draw with your hand resting on the fatty pad at its outer edge. Shoulder lifted and elbow clamped to your side is tiring and awkward, and the tense arm muscles can cause a tremor in the ink line.

Use the pen with a natural firm grip but a light, controlled touch on the paper. You are asking your hand to perform an unfamiliar series of precise movements; therefore it is unrealistic to expect your first effort to be of publishable quality. Practice awhile with the little pen. Learn how to control its marks, and you will have a versatile and responsive tool.

Never dip the pen in the ink bottle. There is a dropper attached to the inside of the cap. Simply touch the dropper to the interior curve (back) of the penpoint. That puts enough ink on it and keeps ink off the penstaff and fingers. Make it a habit to keep the cap sitting loosely on the ink bottle while you are working, to retard the evaporation that can make the ink unusable in a short time. Practice with the pen will also show what happens to lines when the ink begins to dry on the point: the line becomes ragged, fuzzy, and full of skips. If it has dried too much to feed smoothly, touch some more ink to the back of the point and wipe it by gripping it firmly in the penwiper. To some extent ink is a solvent for itself, and the point can be cleaned during use with a little fresh ink. A penpoint that is caked with dried ink will need to soak for a few minutes in pen cleaner solution, preferably the type like liquid soap.

If the pen suddenly begins making a much thicker line, a fiber from the penwiper probably is caught in the slit of the penpoint. It can be dislodged from the back with a thumbnail while the point is gently twisted from side to side. While you are inking a drawing, make a few test marks on a piece of paper every time you wipe and reload the pen. This habitual precaution reduces the risk of bad lines and blots on the drawing. When small errors and blots from lint do happen, stifle the natural impulse to do something about them immediately. For the moment, ignore the blot, make the correction on top of the mistake, and finish all the inking that must be done in that area of the drawing. Then, with the point of an X-Acto knife, delicately scrape out the ink not wanted, working slowly and cleaning the spot with an eraser from time to time. Why one makes a correction first, then scrapes out the bad part, should be obvious; scraping fuzzes the surface of the paper, and no more inking can be done in that spot.

The waterproof india ink used for these illustrations is quick to dry. In conditions of extremely low humidity, it dries so quickly on the penpoint that drawing is interrupted with irritating frequency for the point to be touched with ink, wiped, and reloaded. Indeed, the pen will feed for scarcely more than a few moments under those conditions. As a corrective measure, increase the humidity in the immediate area of the drawing table. An inexpensive vaporizer, the kind meant for infants and sold in most pharmacies, is ideal. Filled with water and plugged in, it emits a small cloud of steam that is instantly taken into the air. Another help is a nearby sink full of water with wet paper towels spread around the rim. If you let them trail in the water, continuous capillary action will keep them wet and constant evaporation will increase the humidity.

Lint-free paper napkins in "cocktail" size--the high-quality multi-layer type--make good disposable penwipers. They come folded once each way; folded once more, the napkin is thick enough to protect your fingers. You can grip and wipe a penpoint time after time at the edge of the folded napkin until the entire perimeter has been used, then trim off the used part with scissors and continue to use its diminishing circumference. Cleansing tissues are second best; their surface releases more lint, an important drawback.

On the subject of lint, discourage any vigorous sweeping and dusting in the room where pencil drawings are exposed and inking is under way. Furthermore, do not allow use nearby of any cleaning product, such as some types of floor wax stripper, that is strong with ammonia. The fumes of the ammonia can be absorbed by the drawing paper, causing the ink line you are applying with such delicacy to spread in a horrifying way. What is more, the effect on the paper can last for several days. Fortunately it does go away in time, and the pencil drawings can then be inked safely. Raising the specter of this unlikely, but possible, scenario is meant to save some illustrator from utter despair in similar circumstances, when a table full of pencil drawings has been exposed to ammonia fumes.

Following is the sequence for inking a pencil drawing. Procedures that are merely mentioned and terms used in the list are explained in detail elsewhere. The need to refer continually to the artifact itself, held in the correct angle of light, cannot be overemphasized.

1. Ink the outline of all views of the piece first; then ink the flake scar outlines, beginning with any small retouch scars and going on to larger scars. Keep checking the scars you are outlining by looking at the artifact; you will be able to improve some scar shapes and resolve occasional ambiguities in the pencil lines. For front and back views, it is a good plan to ink a short distance of outline on one, then ink the corresponding stretch of outline on the other, especially when the edge has many ins and outs. This alternating between front and back views as you draw the outline of the piece makes reversing the fine details of the edge a little easier and serves as an additional check on the accuracy of the outlines.

2. Holding the artifact in correct lighting, consider the general distribution of light and shadow. A flat piece with shallow scars will have little contrast across its entire surface. A thick piece with deeply scooped scars will have more contrast and darker shadowed areas. Draw the direction lines in the flake scars, following the instructions under that listing in chapter 1 and consulting the sample illustrations. Be guided by the lighting on the particular flake scar whose direction lines you are about to draw; do not simply trace with ink the arcs you have penciled there. You drew them to show how the direction lines must curve and where to attach them, but not the degree of shadow that may be required. With the inking comes your chance to give form and character and a three-dimensional quality to your dim graphic statement of data.

3. If the piece has cortex, stipple it last. Well done, it can add solidity and realism. The discussion of cortex in chapter 1 will guide you.

4. When all views of the piece have been inked, the pencil lines can be erased--but with certain restrictions. Erase the pencil lines on the drawings themselves. Erase pencil grid lines between views. Erase parallel guide lines that determined the placement of distal, proximal, and oblique views. To the left and right of the whole group of views, erase all the pencil grid lines, but only the one-fourth inch of the base line and top line that is next to the drawings; the rest of those two are useful at layout time. If you like, you may dim what remains of them, but leave a trace of pencil visible. During all this selective erasing, be careful not to erase the hyphens between the views or any small ticks that mark the points through which a cross section was made. They will be inked after the layouts of drawings are mounted for publication. In case it needs to be said, do not erase the penciled artifact identification below the drawings. Make a notation of each drawing in your records as soon as it is finished; it is useful to keep track of how many pieces and how many views of each piece you have drawn, a way of recording the work already accomplished. Suggestions for record keeping are given in chapter 9.

5. For the present, and until all the drawings bearing that same site designation are finished and layout time has arrived, mount the drawings temporarily, using one or two little tabs of drafting tape, in rows on ordinary sheets of paper--yellow paper is fine--so they can be kept together in a file folder labeled with the full site identification and field season date. If there are scores of drawings to go into the same folder, and many examples of exactly the same tool type, some rough order in how they are stuck on the sheets for filing will be distinctly helpful later. For example, line up all the "straight backed and pointed" together, all the "arch backed," all the "blunt backed," and so on. Tie each group visually with a scrawled bracket and write the tool type for the group, large and plain, on the sheet

they are attached to. That done, the archeologist can tell at a glance if some group of drawings does not represent an adequate sample of that tool category and can select additional specimens from the collection, often the case when material excavated later is being considered for possible illustration. Since the sheets of drawings may be subjected to frequent examination, display to visitors, comparison, reference, general shuffling by a number of people and careless replacement in the folders, it shows foresight to write the site number and field season, as well as the month and year of drawing, across the top of each sheet.

Part 3

The Preparation for Publication

6 Layouts: Composition, Specific Problems, and Reduction

There is a sense of accomplishment when all the required artifacts have been illustrated and the drawings filed away. They may remain in the folders for years, added to from time to time when subsequent expeditions unearth more artifacts at the same sites and samples of those collections need to be illustrated. Eventually all this producing and accumulating of drawings culminates in the signal to make up layouts for publication. This section, on preparing drawings for that final step, provides an introduction to layout composition and the details of procedures to follow. Problems of arrangement and reduction are explored, as are ways of dealing with them. There is also an excursion down the bypath of dummy layouts of artifact drawings as they may be used in contract archeology. Once again, I recommend that you read through the whole section and examine the attendant sample layouts before you open your folders and begin.

Arranging a group of illustrations to form a layout for publication is another of the essential steps between the artifact in your hand and the drawing on the printed page; doing it well will add to the quality of the finished product. The rules in this section have a personal bias and are not universal in all respects, but I offer them as useful guides for assembling artifact drawings into acceptable compositions. They are based on common sense, visual perception, and logic, taking into consideration the reader's convenience and some of the technical requirements of publication.

In practice, layout style is strongly influenced by the preferences and prejudices of the archeologist in charge or the illustrator; but their willful or unknowing disregard of some principles can have deplorable results.

A layout must be more than a group of drawings simply pushed together and pasted down. There is a responsibility to make retrieval of the data as easy as possible for the reader. Be willing to put some thought into layout arrangement. How to impose both rational and visual order on an assortment of elements is the problem. Here is a workable approach and some of the reasoning that underlies it.

General Composition Considerations

Drawings should be set in horizontal rows, to the extent that proximity of varied sizes and shapes allows. Since we read from left to right, the identifying abc's are then in natural order and the eye is not forced to dart around the page in an effort to follow alphabetical references in the text and caption. When drawings are dotted loosely about the layout area, without the suggestion of order that horizontal rows impart, they produce an unstable floating effect.

Trying to save space by fitting as many drawings as possible into a page does not justify one of the most jarring of layout faults: turning some illustrations to positions different from that dictated by tool type and individual features, a change that violates the correct orientation in which they were interpreted, lighted, and drawn. The violation not only shatters the visual unity of the page, it also introduces an obstacle to the reader's perception and comprehension.

In one style of well-knit layout that makes economical use of space, drawings touch the top and side margins; the bottoms of the views on the final row form a straight line so that the abc's below them sit on the bottom boundary of the layout. The interior rows of drawings also are neatly horizontal in their general trend. However, a few may step up or down a bit to clear a tall piece on a lower row or duck below a piece on the top row where, with distal ends of illustrations against the top margin, their bottoms form an irregular line, determined by the lengths of the artifacts. This step up or step down is helpful near pieces whose distal or proximal views would otherwise leave excessive empty space around them (fig. 31; fig. 33). If the maneuvre is used with restraint, so as not to destroy the orderly effect of horizontal rows, the drawings forced to step up or down help to stablize and lock together the rows of irregular shapes.

Put the smallest, lightest-looking drawings at the top of the layout and the largest, heaviest-looking ones in the bottom row (fig. 13; fig. 17). Arrange the other pieces in the middle rows so that, in effect, the layout grades downward gradually from small, light, and simple to large, heavy, and complex. This puts the weight at the bottom and avoids a top-heavy layout. In general, an artifact drawing with few flake scars does not look as heavy, and certainly not as complex, as a piece of similar size and shape exhibiting many flake scars.

A "weak" artifact (i.e., pale, fragile, very small), rightly assigned to the top row (fig. 26, *c*, for example), should avoid a corner if possible. A little strength at corners is desirable, but not so much that the arrangement appears to be staked out emphatically at those points.

Try to keep together in series all the specimens of one tool type category, size range permitting. A tiny thumbnail scraper and a large carinated endscraper, for example, are too different in size to be mounted side by side, even though they are both scrapers. The small piece should be high up on the layout, with other light weights, and the large scraper much lower down because of its weighty appearance. In such a case, only extreme disparity in size forces the separation of examples of a single tool type. Above all, do not engage in random mixing of drawings in complete disregard of their classifications. The reader should not have to grope through such identifications as "*b*, *g*, *s*, *x*: *endscrapers*; *d*, *f*, *y*: *retouched pieces*."

Even if an archeologist's harried response to a query about composition is "I don't care--just get them in there some way," the illustrator still has a professional responsibility to produce orderly layouts, in justice to the effort that has gone into the drawings as well as their importance as extensions of the archeologist's text. Carefully assembled layouts of good drawings add a quality known as class.

A tool category whose members vary slightly in size but not in gross shape, such as projectile points or backed bladelets, should have those sizes mixed, not stairstepped, on the layout. This pronouncement flies in the face of what seems common practice, especially in New World archeology. Nevertheless, a row of such pieces, their sizes evenly graduated, appears disconnected from the other drawings and visually "snaps off" from the rest of the layout. Each piece is granted its full value and individual importance when the sizes of the look-alikes are mixed.

If a site has generated enough drawings of assorted tool types to make several layouts, do not split a tool category between two pages. When, for example, one layout area will accommodate all the perforators, points, geometrics, and scrapers but has room for only a few of the burins, do not split the burins between that layout and another one. In this example a better plan would be a short layout of perforators, points, and geometrics; then make another layout with burins and scrapers. If there is also a "loner" (a single piece, such as one notch, one truncation, one denticulate), put it with the layout that needs it to fill out the composition, unless the loner's type is more compatible with a particular layout's tools. For instance, in this hypothetical case a denticulate probably should go on the layout that includes the scrapers.

In deciding how to divide all the tool types of one site to make several layouts, consider "kinship." As an example of logical kinship, microburins would certainly belong on the layout with geometrics. All of the many varieties of backed pieces, such as "straight backed," "arch backed," "partially backed," and so forth, obviously belong together. Mounted with them might be the "backed and truncated" and "perforators." This does not mean that a layout should be limited to "kin"; only that kinship can help make logical groupings.

A site, or even a subdivision of a site, may have several hundred drawings of perhaps a hundred pieces and require half a dozen or a dozen layouts, depending on the general size of the artifacts. An assemblage consisting primarily of scores of bladelets and microblades will tempt one to crowd as many as possible into the full layout allowance. There is reason to resist the temptation: So many little pieces on one layout would probably run through not only one alphabet of identifying abc's, but also most of another alphabet of double a, double b, and so on. In that case more caption space would be needed for the descriptive matter that must refer to those abc's and name the tool types of all pieces. And this is extra space you do not have; before a desired percentage of reduction could be used to determine the measurements for the layout area itself, an arbitrary allowance of space for captions was first subtracted from the length of the effective page (discussed in the sizing of layouts). Rather than create this problem of not enough room to identify the artifacts in a layout full of microlithic tools, divide the drawings in a logical way for two or more shorter layouts.

Avoid long white "rivers" running down the layout. A river is a noticeable band of white that occurs when the spaces between artifacts accidentally fall one below another, row after row. A page will appear to break apart along that white band. Swapping the positions of one or two pairs of pieces usually cures this defect without disturbing the logic of the arrangement.

Concern with such details as eliminating rivers is part of the process of "fine-tuning" a layout. Careful measurements, level placement, and overall neatness are basic and expected; but some thoughtful squinting at the arrangement through half-closed eyes can suggest small adjustments in position or spacing or alignment that will further improve its appearance.

Specific Layout Problems

When layouts of illustrations are intended for publication in a journal, study some back issues of the periodical, not only to get the measurements of its effective page that will determine layout proportions, but to discover the characteristics of its format. See if it publishes full-page figures, or pages that combine short layouts with text, or both. Note especially if there is a strong tendency toward drastic reduction of illustrations regardless of how they suffer for it, a significant point to ponder if the artifacts you have drawn are primarily little backed bladelets, for example. A generous layout, thickly set with many small pieces, would be asking for trouble. Such a case calls for smaller layouts that will not force the journal into severe reduction simply to get them on the page. Plan your layouts so they will not cause problems when the manuscript they illustrate is submitted to the journal for publication.

One-Column Layouts

If the journal's page is divided into two columns and it publishes figures that are one column wide as well as figures that use the full width of the effective page, use that fact to solve the problem of composing a satisfactory layout when a site with scanty material provides only a few drawings of very small tools. A full-size layout area with four or five tiny pieces in a sea of white is undesirable in effect and wastes valuable space. Measure the column width and use it to figure a small layout size, possibly one-third to one-half the column's length. Think again about reduction; it may be best to leave extremely small artifacts, such as "bird" points, bone needles, and ostrich eggshell beads, at natural size. Whatever the decision, remember that every layout, whether reduced or not, must have a scale in it.

If possible, investigate the journal's format even before drawing the artifacts that are to be published. For one-column layouts in particular, some time can be saved and a minor nuisance avoided if the multiple views of one artifact that can almost fill the column width are drawn to fit it precisely. A small adjustment of the space between the views while they are being drawn makes the difference.

Take the case of an arrow point that needs a front, back, and profile drawn. Rough estimating shows there is room across the column for two "widths" of the artifact and a "thickness," with a little space left over but not enough to set the drawing of another artifact beside it. Draw the front; rule the base line on which all the views will sit and a vertical touching the left edge of the front view. From the spot where the vertical touches the base line, measure off the column width and rule another vertical there. Draw the back view with its right edge touching that vertical. Then draw the profile between the views of front and back. The whole unit of multiple views of the artifact will fit precisely into its layout, with no need to cut the views apart, spread them to take up the extra bit of space, and realign them. This is a reminder that the artifact drawing is not the end product, and that looking ahead can prevent a headache later.

Core Layouts

Drawings of cores, ideally exhibited in their own layouts and not mixed with tools, are inherently difficult to arrange in a satisfactory manner. Two or three cores of widely differing sizes are a real problem. Eight or ten cores, all medium-sized lumps, are awkward and monotonous. A balanced, well-knit layout of cores is rare and a triumph (fig. 56, for example). Luck of the draw gets most of the credit or the blame, but a few helpful maneuvres are possible.

Part of the difficulty lies in the indivisible groups of three views required by most cores. Unless the cores are rather small, there isn't room for two "threes" on one row, and yet one "three" leaves too much space if it is placed to touch either the left or the right boundary of the layout area. You are fortunate if the assemblage has a few cores (usually single platform) that can be illustrated adequately with only two views, providing some narrower elements to be paired with "threes" and thus fill rows comfortably the full width of the layout (fig. 57). When you are faced with a number of "threes," none of which can be paired on one row, a possible plan is to arrange them to touch the side boundaries of the layout, alternately left and right, as they go down the page (fig. 58). If this plan is followed, the appearance can be improved by placing the "rows" quite close together, tightening the layout from the bottom. When cores awaiting illustration suggest that this problem is ahead, draw their multiple views with somewhat more than the customary space between them, perhaps half an inch instead of one-eighth or one-fourth. For help with this estimating, see the discussion of rough layout forms later in this chapter. Reminder: the obligatory scale can help fill a hole in a core layout, too.

Globular cores often need four views, and so does an occasional pyramidal core. The sample of cores from a single site subdivision very seldom contains more than one specimen that requires four views, but it often presents the same problem as a "three" in neither filling a row nor leaving room for another core. "Fours" can, in desperation, be centered alone as a "row" that touches neither left nor right boundaryes. It can even provide a little variety in an otherwise monotonous layout.

Rather large cores--say, two inches across--will come close to spanning an average layout with their three views (fig. 58; fig. 59; fig. 60). The spaces between are the variables. Follow the efficient system of drawing to fit the space, as detailed in the discussion of one-column layouts. Better that than finding the three views must be cut apart and spread, or moved closer together, with the attendant nuisance of realigning them--and then finding a paper edge exactly where you want to ink the hyphens at the finish.

Biface Layouts

Large bifaces, from the illustrator's point of view, are both challenging to draw and beastly to arrange in layouts. Front, profile, and back are usual (fig. 47; fig. 48); a cross section of a tip is at times desirable, but some archeologists insist on a cross section through the wide part or through backing. This means another shape has to be drawn below, and lined up with, the primary view. It adds the further complication of imbalance to that of their awkward size. In that case you must settle for less that ideal composition.

An assemblage consisting primarily of bifaces ("handaxes") should be drawn and mounted with their special problems influencing your decisions. A number of the artifacts may be even larger in the flesh than the entire page on which they are to be published. The smallest one may be as big as the most massive tool in an ordinary assortment. Unfortunately, drastic reduction of their illustrations is the only way to get them on the page of an average-size book.

Reducing the drawings of bifaces to two-thirds of natural size is common. This amount of reduction will usually allow space for two artifacts to a layout, except in the case of the largest ones, whose three (or more) views will take up a page. For them, make "horizontal" layouts, meant to be run broadside and viewed by turning the book sideways, using the *length* of the effective page to figure the *width* of the layout. Subtract from the width of the effective page measurement an allowance for a caption (perhaps one-fourth inch); what remains is used to figure the vertical "length" of the horizontal-shaped layout. Note that less space is needed for captions that are to be read with the book turned sideways because the type will be running across the length of the effective page instead of the width. Also, only one artifact on the page means less descriptive matter is required if merely figure number, site designation, and tool type are all the identification planned.

Ideal for illustrations of bifaces is the extra large format of some European books, such as *Typologie du paléolithique ancien et moyen* (1961), written (and illustrated in part) by F. Bordes, in which Pierre Laurent's superb drawings of bifaces are allowed to display the artifacts at natural size.

Forms for Rough Layouts

A layout form is a practical aid in working out rough arrangements of illustrations. It can be simply a sheet of typewriter paper with a rectangle, representing the layout area, carefully measured and ruled in ink. The kind of pen or ink is unimportant, but the ink line should be strong enough to be seen, at least dimly, through the paper used for artifact drawings. The measurements are those that will be used when the original drawings are mounted on rigid showcard board in finished layouts. As a precaution, and for future reference, write in a margin of this form the measurements of the rectangle, the percentage of reduction for which it was figured, and the name of the journal or book in which the layout is to be published. Then copies of this one sheet can be run off as needed.

The layout form is useful to both the illustrator and the archeologist. A copy of it at hand provides a quick way to gauge the possibility of fitting the multiple views of a large artifact into the pescribed measurements. Laying artifacts around on the sheet, with informed guesses as to the space needed for their additional views, can give an estimate of what pieces might make one layout and even show that one more artifact of a certain size would fill a hole in the arrangement. Furthermore, it gives a rough approximation of a layout on those occasions when the archeologist wants only a few layouts of a representative sample of tools to accompany an article and the complete artifact collection has not yet been illustrated.

For the archeologist whose illustrator is not sufficiently familiar with typology and the logical association of tool types, the inexpensive copies of the layout form supply an efficient way to handle the grouping of tool types for mounting. Scores of finished drawings (*do not trim them* at this stage) can be shifted around into tentative rough arrangements on a series of the forms and stuck in place with bits of drafting tape. These mock-ups will indicate the desired division of tool types; the illustrator can make appropriate adjustments within the groups that will turn them into acceptable compositions when the drawings are transferred to the layout boards.

A note of caution: be aware that photocopies of the layout form often do not exactly match the measurements of the original. The amount of difference varies with the machine on which the copies are produced. Be sure to check the measurements of the copies; a too-tight tentative arrangement on a slightly enlarged form will not fit the precisely measured rectangle into which the drawings will be mounted as finished layouts. The error can be compounded if a copy of the form is photocopied for an additional supply. Keep the original of the layout form on file and copy it when extras are needed. Also, be aware that photocopies of artifact drawings, too, can differ in size from the originals, a fact of some significance in the case of dummy layouts made up of such copies. Spinning out the possibility of error, a photocopy record of a dummy layout, made of photocopied drawings mounted on a copy of a photocopied form, may be far from the original measurements--in no way to be trusted for, say, matching measurements of spaces between drawings in the final layouts.

A rule that may seem overstated but is not a bad one to follow: Never trust measurements taken by anyone except yourself--and then only those you take from an original, using the same ruler every time. It is unsettling to discover that measurements made with an assortment of rulers can vary by several sixteenths of an inch in less than a foot of distance.

Effect of Reduction

In some respects the ideal way to publish artifact drawings would be reproduction at "natural size." From the artist's viewpoint, it is desirable for drawings to meet the public gaze exactly as drawn. The inhibiting question, Will this line (or area) survive reduction? could be forgotten, and whatever delicacy of line was appropriate for elaborate flaking would be feasible. From the archeologist's viewpoint, "natural size" would match in graphic image the description of the artifacts' attributes in his text. For the reader of the published work, "natural size" would be the next best thing to examining the artifacts, useful for comparison with his own collections, since it is unlikely the reader would ever have access to the artifacts themselves.

However, the costs of publishing and the size of the sample considered representative of a particular assemblage of pieces force a more realistic, though less satisfying, treatment. Photographic reduction of the drawings makes it possible to display more artifacts in a given amount of page space. If that were the only consideration, a page could be packed with tiny shapes; but the effects of drastic reduction on the data that such drawings embody would be destructive and unacceptable.

Reduction by 10 percent is not really damaging; 20 percent reduction is near the maximum amount that the average run of artifact sizes can survive successfully. Certainly, microlithic assemblages cannot tolerate more than 20 percent reduction without excessive loss of detail; reduction of no more than 10 percent is preferable.

At the other end of the size range, large bifaces also have problems associated with reduction. When a collection includes a number of bifaces, it is usual practice to reduce their drawings by one-third, for the obvious practical necessity of forcing one or two to fit on each page. Unfortunately, examples of the most extreme reduction, to merely one-fourth or one-fifth of natural size, are not hard to find. Not only is there a degree of reduction beyond which even gross detail is lost, but "amygdaloid" or "cordiform" images the size of a thimble are simply not perceived at their life size, although the reader's mind may register the scale and the fact that reduction is severe. An archeologist once compared his excavated material with some of those miniature images. He decided that his artifacts ran much larger, and that resemblance therefore was superficial and negligible. Only somewhat later did it occur to him that he had been led to a wrong conclusion by the extremely small size of the illustrations themselves. Despite his awareness of the scale, the artifacts had lost for him even the suggestion of their true size when they were pulled too far from it by reduction.

When the amount of reduction to anticipate is not known, adjust the style of drawing bifaces in the expectation of reduction to, say, two-thirds natural size. When everything you put on paper is one-third smaller, lines of spiderweb thickness will break or disappear. Ink biface drawings with a firm hand. Lines after reduction are thinner, of course, but they are also closer together, tending to produce a darker effect in general. Therefore use fewer direction lines to maintain a more open look to the flake scars. The very smallest retouch scars on edges will have a tendency to close and black out if their tiny areas of white are made even smaller by unnecessary direction lines. Compensating for the effects of drastic reduction while you ink the pencil drawing is a way to prevent, to some extent, the loss of retouch detail.

Transforming the layout of original ink drawings into the picture printed on a page can produce distressing changes. A type of photographic print, a "reproduction stat," with a film negative gives results closest to the original drawings. At best every process, after the illustrator's work is done, takes away something, even in the hands of skilled craftsmen. Some control over the quality of at least the first step is possible if the illustrator or the archeologist will select a good photostat service. For leads, and informed professional judgments on relative merit, inquire among commercial artists, especially those who work for advertising agencies. When a photostat company with a reputation for painstaking work is located, have a talk with the person who will be responsible for making the "repro stats" of your drawings. Explain that all the marks in the drawings have scientific meaning--that these are not merely sketches. Emphasize the importance of picking up all the fine lines, but without closing up the small scars of retouch and backing. A skilled photostat man is a valuable ally, and establishing a good working relationship with him is beneficial. His knowledge of his field, together with an awareness of the special needs and problems of yours, can determine the quality of repro stats and how closely they match your original drawings.

As a general precaution, repro stats of the artifact drawings, not the original art, should be sent with the manuscript to the publisher. A stack of layouts of these drawings--say, enough to illustrate a book--can represent such an investment of time and effort and money (in many cases years of work) that they are literally irreplaceable. The artifacts from which they were drawn may have been returned to the other side of the world by the time a book goes to press. And that spot on the other side of the world may be engulfed in political upheaval. Therefore artifact drawings are not to be regarded as "drafting" or classed with charts, graphs, and simple site maps.

7 Dummy Layouts of Drawings for Contract Archeology

"Dummy layouts," as the term is used in lithic illustration, are exact mock-ups of what will be the finished layouts of the artifact drawings--exact in measurements and arrangement, but with the drawings themselves in an unfinished state. The mock-ups are made the same size as the intended layouts--not the reduced size that will appear on the published page.

In contract archeology, federal and state agencies often permit dummy layouts to accompany the first submitted draft form of a project report. In fact, the more experienced and knowledgeable contract administrators prefer this because of the unique circumstances of producing the illustrations. It is a needless risk to send original art by mail when photocopies can show what work has been done and give some idea of the quality to expect in the final version of the report. Of course, the individual administrator would base his decision on his own experience with the job performance of the people involved in the project and a reasonable expectation of acceptable quality in the finished drawings. Also, the professionalism of dependability in meeting publication deadlines would influence his agreement to the dummy layout system.

This intermediate step in the artwork makes more efficient use of the time available for drawing. In that final push to meet the deadline, dummy layouts work to the benefit of the archeologist as well as the illustrator by meshing the labors of both.

Assembled artifacts have individual identities that require their locations within layouts to be tagged for reference in the text. Dummy layouts settle the arrangement of the separate drawings earlier in the time span of the job than is possible if illustrations are finished first, then the layouts made. This change in the scheduled order of the steps gives the archeologist more time to write the captions, insert the necessary references to individual artifacts in the text (positions of the pieces and identifying abc's are now final), and assign figure numbers. Consequently the rough draft can be clean typed under a little less pressure to meet the date for submitting the report to the contracting agency.

Meanwhile, with the dummy layouts in the hands of the archeologist for integration with the text, the illustrator can begin to ink the original penciled drawings. Mounting them in finished form is then a matter of copying the arrangements already decided in the dummy layouts.

To use the system just described, make the normal detailed pencil drawings of all the artifacts that have to be illustrated for the report. Use a 2H drawing pencil in this case; although its softer lead needs sharpening more often than the harder 5H otherwise recommended for artifact drawing, blacker lines are needed here. Then, before inking any of the pencil drawings, make inexpensive same-size photocopies. Any copier that will pick up the pencil lines adequately will serve--even coin-operated machines in post offices and drugstores--but test several machines to find one that works well. The prints need not be perfect replicas, but they should register the outlines of the pieces and enough of the detail to make tool types recognizable.

If you have followed the drawing instructions in this manual, each artifact has been drawn on its own piece of paper. Before photocopying, attach these separate pieces to sheets of typewriter paper, as many as will fit on each sheet without overlapping. Two or three tabs of drafting tape will hold them to the sheets and make them easier to handle for copying. (*Do not trim the pencil drawings*; that step comes later, after they are inked.) How they are arranged on the sheets does not matter at this stage; in fact, if both standard and legal size copies are the same price, stick the loose drawings on legal size paper and copy more for the money.

Put away the originals, for the time being, and cut out the photocopy prints of the penciled illustrations exactly as if they were finished drawings, according to the specific instructions on trimming in the next chapter. Next (detailed later) mount the photocopies in correctly sized and carefully planned layouts, but with this difference: use sheets of standard typewriter paper, instead of rigid showcard board, for the dummy layouts so they can be inserted conveniently in the draft of the report. Remember to put a scale in every layout and freehand the abc's in approximately correct position with a pen.

On a separate piece of paper for each layout, write a "key"--a memo of its contents. If you have followed the instructions on penciling artifacts, the full identity of each piece is on its drawing, for reference in making up such keys. The key's heading should be complete formal site designation, usually a combination of numerals and letters; the name by which the site is known, such as "Bir Kiseiba" or "Hog Creek"; and the year of the field season that produced the material illustrated, for example, "1978 Season." Then, in rows that correspond to the arrangement of the layout, write the individual identification of each piece and its tool type, with an alphabet letter below each identity to match that on the dummy layout. One-fourth of a sheet of typewriter paper is a good size for the key because of this next essential step.

Photocopy each completed dummy layout on legal-size paper, with its key placed beside it on the machine so they will be copied on the same print. The key may be turned any way that does not cover any of the drawings. This combined photocopy of layout and key is simply a record that the artist must keep in order to duplicate the arrangements when the time comes to make up the finished layouts of the original drawings themselves; the dummy layouts will have been sent away with the preliminary draft of the project report. Photocopying the key onto the same print with its layout is a safeguard against mix-ups.

The key itself is for the benefit of the archeologist who must write the caption. It can save a search through a collection of artifacts to locate and reexamine a particular piece whose morphology is ambiguous. No more muttering, "Now what did I decide to call that?"

The admonition to tape a copy of the key to the back of the mounting board is important enough to repeat. Often, layouts of original drawings are stored for years before they are published--and for years after; the artifact collections they represent may be unavailable for reexamination when the time comes to publish--or republish. Furthermore, the archeologist who originally classified the artifacts may be long gone. Never trust the identification of artifacts or drawings to anyone's memory.

8 The Mechanics of Making the Finished Layout

Making layouts of the drawings is the last big task for the illustrator on the way to publication. You have been told how to approach the project, briefed on problems and solutions, and forewarned against a few perils. Now, here are the mechanics of the job.

First, find out the measurements of the "effective page" of the anticipated publication--that area of the page, within the margins, available for text or illustrations. If it is the archeologist's own book, he will have to supply the information on effective page size and the amount of reduction he is considering. The illustrator may speak persuasively of the advantages of minimal reduction, but never argue with the archeologist.

If the drawings are to be published at natural size (strongly recommended for microlithic assemblages), subtract from the length of the effective page an allowance of space that will be used for the caption, sometimes called "the cutline" or "the legend." This is the printed matter that gives the figure number and identifies the drawings in the layout. In practice, this allowance is often sheer guesswork. If the archeologist knows that he will need only one line of type (the figure number and something like "Tools from Site 42, Level 6"), one-fourth inch is enough. If more complete identification of the individual artifacts by alphabetical designation and tool type is planned, more lines of type will be needed; therefore a larger allowance of space for the caption must be subtracted--say, one-half or five-eighths of an inch.

To repeat, if the drawings are to be published at natural size, what is left after you have subtracted the caption allowance from the effective page's length is the size you will make your layouts. It is the size of the rectangle you will rule with non-photo blue pencil ("drop-out blue" pencil) on rigid showcard or illustration board and within which you will arrange your drawings, their abc's, and a scale.

If layouts are to be made up larger than for publishing at natural size, meaning that photographic reduction will be required to make them fit the available area of the page, the measurements of the layout's width and length will have to be greater, but they must maintain the same proportions. If you can work it out mathematically, fine; quicker and less painful for most is a "proportion wheel"--a sort of circular slide rule that is nonintimidating, can be mastered in five minutes, fills the user with confidence, and can be purchased for a few dollars at most art supply stores. Two thin plastic disks--one a little larger than the other, both smaller than a saucer--have been fastened together through their centers with a brad that allows them to be turned. Around the perimeters of both are many little ticks and numbers that represent inches and fractions thereof. By turning the uppermost disk, one can line up certain ticks on the two perimeters. Match up a width you have with a width you want; then find the length you have, and there, automatically matched with it, will be the length you need to know. Some models have a small window that reveals the "percentage of original size" the new measurements represent. That feature is especially useful.

If you need to visualize several possible layout sizes, a simple ruled diagram that you can concoct quickly on a sheet of paper shows all. It is a proportion scale of right-angled lines intersected by an extended diagonal that runs through the corners of a known rectangle, a handy device for testing a choice of enlargements (or reductions) that must keep the same proportions. Here is how you can make one in a few minutes.

In the lower left portion of the sheet of paper, draw a rectangle the exact size of the effective page. Rule a line across the bottom of it to fence off the part that will be occupied by the caption. That strip has been "subtracted"; ignore it from now on. Extend the "fence" line (now the southern boundary of the rectangle) eastward to the right edge of the sheet. Extend the line that is the western boundary of the rectangle northward to the top edge of the sheet. Rule a long diagonal line that starts at the southwest corner of the rectangle and travels right through and on out the northeast corner, to end where it will, at the edge of the sheet. Those southern and western boundary lines that you extended will always form the bottom side and left side of any layout size you select. Starting at the southwest corner, measure along the southern boundary line what you want the width of your layout to be, and there put a dot. From the dot, run an exact vertical line northward until it crosses the long diagonal line. That crossing point is important; it will be the location of the top right corner of the layout when you have supplied the final (top) side by drawing a line that simultaneously forms a right angle with the extended western line and crosses the long diagonal exactly at its intersection with the other line. Now you have a new, larger rectangle that retains the same proportions as the original one. (For a similar method and a diagram, see Zweifel 1961.)

Mount drawings on a rigid material only, either illustration board or 14-ply showcard board with a smooth "hot press" surface, not the "cold press" kind that has a surface like blotting paper. Springy, flexible poster board is not suitable. Hot press illustration board is available in sheets 30" x 40" which at this time cost $2.65. Perfectly satisfactory and much cheaper is 14-ply hot press showcard board, which comes in sheets 28" x 44" and at present costs $1.65.

The big sheets of mounting board must be cut into pieces, each of which will be used for a single layout of drawings. Not every illustrator has access to a paper cutter with a blade long enough for that first cut--or even the second. Most comprehensive art supply stores have an extra-large paper cutter. Some shops charge a fee (at so much per cut) if they do the cutting for you; others do not charge for the service. Often, however, the customer is allowed to use the cutter. Even one cut across each sheet makes a more manageable size for carrying; further cutting to final size can be done on a small cutter.

When the size of the layout rectangle has been decided, it is good planning to figure the most economical cut of the big sheet of board. Looking ahead to future work, it is possible to select a cut that will give a uniform size for each mounting board. The size chosen should accommodate a reasonable range of layout sizes and proportions and still leave plenty of margin--at least one and one-half inches, preferably more. For example, the 28" x 44" size cuts exactly into eight pieces, each one 11" x 14", with no waste. That size, 11" x 14", is generous enough for most artifact layouts.

There are other advantages, besides economical cut, to be gained by adopting a single size where artifact illustrating is a continuing activity, such as the archeology department of a university. Uniform size makes storing the layouts easier and neater. Another consideration is the sizes in which brown envelopes are available. This 11" x 14" cut will fit a standard envelope, 12" x 15½", to protect and segregate, by individual site subdivisions, up to eight or nine layouts. Shuffling through mounted drawings again and again in search of a certain site can be destructive in a short time. Store the mounted layouts in brown envelopes with the site written very large on each envelope--and store them flat to avoid warping. If a supply of the large sheets of board is kept on hand, store them flat also. Cut small batches of the size needed for an anticipated series of layouts; need for a large piece of the board can arise.

Use only a "non-photo blue" pencil to draw the rectangle within which the illustrations are to be arranged. This special pencil will not be picked up by the camera and therefore marks made with it need not be erased. Carefully rule a layout area of the measurements you have already figured, leaving a minimum 1½" top margin and centering the rectangle from side to side on the board. The bottom margin is determined by the length of the layout. Rule the rectangle full layout size even though arrangements of drawings often fill less than the full length.

Now spread out your folder of drawings and rapidly scrawl the site number and tool type abbreviation on the back of each, lightly in pencil, keeping the identification within the area occupied by the drawing on the front so that the writing will not be lost when the drawing is trimmed later. Adopting this little precaution as a routine part of layout preparation has been justified many times. When layouts are split up, sampled from, or recombined into different formats for other publications (often years later), they have to be removed from the boards for remounting. Loose, trimmed drawings are too vulnerable to loss of identity that renders them useless; site number and tool type on their backs prevents that loss.

Every layout must have a scale in it, whether or not it is to be reduced. To rid yourself of the nuisance of drawing scales every time layouts are made up, draft a pageful of them--very neat, very precise--with at least an inch between them in the row and between rows. Keep this master copy filed away. When layouts are to be made, run off a good clean photocopy to cut up and use in the layouts.

A scale should be as simple and as unobtrusive as possible--no checkered effects to snag the gaze and compete with the illustrations for attention. It should be quietly present in case it is needed for reference, doing what it can in the meantime to help the composition by filling a hole or cutting a river. A double-zero penpoint for a line, and ticks one centimeter apart, is a personal choice.

Sort the drawings into tool types and decide which types might be grouped logically on the same layout. If the site has so many drawings that three or four layouts will be needed, either prepare that many layout boards to work on simultaneously or use layout forms (described in chapter 6, on layout composition) for figuring out the arrangements. Do not trim the drawings yet. Following the advice on composition, move the drawings around until you are reasonably sure they will fit the space and not grossly violate the principles of order, logic, and balance discussed at the beginning of the section on layouts.

Now you are ready to trim the drawings, for one layout at a time; however, there are factors that determine how the job should be done. In the first place, excess drawing paper around an artifact's group of views must be removed to keep it from interfering with neighboring illustrations on the layout. Second, enough margin must be left outside the drawings so that the tape used to stick them to the board can grip the edge of the paper but never touch the drawings themselves. A third factor is the need to guard against ever losing the exact orientation of a one-view artifact and also to prevent accidentally mounting any piece upside-down. The fourth factor looks ahead to the placement of the abc's below the drawings. A fifth consideration is the need to bridge the identification gap between the mounted, trimmed (and therefore nameless) drawings and the production of a layout key that the illustrator must provide for the archeologist to use when he writes the caption for the layout.

Coping with all these factors at the same time calls for a trimming in one piece, like an apple paring, from around the drawings of a single artifact, with the cutting done this way:

> First, make a *straight* cut across the bottom of the group of views (or single view) of the artifact, one-sixteenth of an inch below the base line and parallel to it; stop about a quarter-inch past the spread of the drawing. If a proximal view is present, or any view or symbol that extends below the base line, consider it to be the bottom and "base line" of the group and make the straight cut below it instead. The bit of base line you were told to leave when you erased the rest of the pencil lines helps guide the scissors. The one-sixteenth-inch margin allows barely enough paper for the tape to grip, but it is necessary to get the edge of the drawing paper out of the way of the letter that will be supplied by dry transfer type, close below each illustration. The *straight* cut will forevermore indicate the bottom of the drawing; and, since that straight is horizontal, the exact orientation of the artifact is preserved.
>
> Now turn the scissors away from the straight cut and trim outside the rest of the group, about a quarter-inch from the drawings, in a rounding effect, as you roughly follow the general shape of the piece (if a single view) or the group of views (if multiple). Trim a little more, here and there as needed, for fit with neighbors in the layout.
>
> As you trim around the drawing(s) of each artifact in turn, replace it in the tentative arrangement already worked out and lay its

trimming on the table in a position, relative to other trimmings, that repeats the arrangement of the drawings themselves. Each trimming carries the full identification of its artifact. When drawings are moved around in an effort to compose a satisfactory layout, move the trimmings also to corresponding positions. When the arrangement has been settled and the drawings taped in place, the positions of the trimmings and the artifact data on them will supply the information for writing the key.

The "hole" in the middle of the intact trimming provides a quick way to confirm the identity of a drawing already taped down. Repeated rearrangement in the course of making a layout can cause that feeling of "Did I swap the trimmings when I swapped the drawings?" It is easy to remove that uncertainty. Slip a trimming over a mounted illustration to check size and shape against the hole rather than cutting the drawing loose from the board to verify the identity written on the back.

For the sake of clarity, uniformity of format, and professional finish, adopt a formula for consistent placement of abc's on layouts. The following personal formula is an example:

Center the letter

-- below a one-view artifact (fig. 15, *d*).
-- below and between the views of a two-view artifact if both sit on the base line (fig. 16, *c*); below the main view if the other does not reach the base line (for example, a partial view of an upper portion: fig. 40, *b*).
-- below the middle view of a three-view artifact (fig. 40, *d*).
-- below and between the second and third views of a four-view artifact (fig. 38, *l*).
-- below an artifact with a butt, proximal, or cross section view that is drawn below the base line, in conformity with the

above rules that are applicable, but as if the bottom of the low-placed view were the base line (fig. 54, *b*).

Now, briefed on possible problems and forearmed with the reasoned suggestions in this section, you are ready to adjust the trimmed drawings and a scale into precise placement within the rectangle you ruled with non-photo blue pencil. Line up the illustrations so that the outlines of the outermost drawings against left, right, and top boundaries barely touch the rectangle's outline. Their margins of drawing paper and the tape that holds them down will extend over the blue lines, but that is all right. At the same time, check that the drawings are not tilted to one side or the other. Near the spots where letters will go, make sure no drawing on a lower row crowds too close.

When illustrations for a single layout need the full length of the area, the final row of drawings must have enough space below them for their identifying letters. For example, setting that last row of drawings six millimeters above the bottom blue line boundary will leave room for the letters if the 14-point size of dry transfer type is the one to be used. Consider the bottom boundary of the layout area as the line on which to set the lowercase letters; the tails of g's, p's, and such, will extend, as they normally would, below the line.

Do a lot of squinting through half-closed eyes. Consider visual weight distribution. Test some minor changes in the spacing of the rows. See where the scale will do the most good. Maybe there is a "weak" piece against a boundary; maybe another piece could swap places with it. And remember to swap their trimmings.

When you decide you have arrived at a logical, orderly arrangement and cannot improve on the composition, stick each drawing precisely in place with temporary snippets of drafting tape, about one-eighth inch wide: put one little piece where the identifying letter will be (to remind you not to stick the permanent tape in that spot) and two or three more bits of drafting tape around the edges to hold the drawing until the "invisible" frosted tape can be used for the final sticking down. Incidentally, you can use the little tabs of drafting tape more than once. Stick them along the very edge of the table, or other handy free edge, as you replace them with frosted tape.

This is a good time to write up the key for that particular layout; the order and positions of its trimmings are vulnerable to accidental disarrangement. When their information has been copied in matching sequence, as discussed in the section on dummy layouts, the trimmings can be discarded.

The importance of the next move cannot be overemphasized: the complete, formal site designation must be put on every layout. With a hard-lead pencil--not a pen or a pencil that can smear--clearly print the whole site number along the bottom edge of the layout board, well away from the area occupied by the drawings. Add the site's name (such as "Tabun" or "Hog Creek") and if possible the year of the field season that produced the artifacts on that layout ("1978 Season"). Do not number the layout; the archeologist will determine the figure number when layouts of drawings, along with pieces of drafting such as maps and graphs, are integrated with his manuscript.

For the next step--which involves the use of Scotch brand Magic Transparent Tape, preferably one-half inch wide--hands must be washed with soap and dried, and then the fingertips and nails recleaned with a piece of the tape to remove any lint or fibers. Also, clean the blades of the scissors with some of the tape. During the whole sticking-down process, keep your hands this clean.

Cut with scissors--do not tear with the serrated blade on the dispenser--small pieces of the frosted tape and affix them lengthwise around the edges of the drawings to fasten them, almost invisibly, to the mounting board. A serrated tear exposes more "sticky" than a smooth cut and is therefore more likely to catch airborne dust and dark fabric fibers.

If you are thinking of using paste or rubber cement--don't. Often, individual drawings must be removed from layouts and combined with others in different formats for publication. After a period in storage, rubber cement can produce a brown stain; glued or pasted drawings may either pop off the board spontaneously or be impossible to remove intact. It is quick and neat to free a Scotch-taped drawing by simply cutting through the tabs of tape, right against the drawing paper's edge, with the very point of an X-Acto knife or razor blade. Then mount the drawing in the new layout with fresh tabs of the tape. Do not remove the remnants of tape still adhering to the margin of the drawing paper, because some of the adhesive may cling to the paper, collecting dirt and lint. If for some reason bits of tape must be removed, a fresh piece of tape can be dabbed over the area to pick up any remaining traces of the adhesive.

Dirt and fibers must be kept off the roll of tape also. Whether it is the end of the project or simply the end of the working day, store the roll in a closed sack or plastic sandwich bag; do not toss it in a drawer or leave it in general use. The tape used on layouts must be clean; dark fibers can show up on the repro stats. It is even advisable not to wear fuzzy dark clothing on days when layouts are to be taped. The precaution will not seem farfetched when you have had to remove and replace some pieces of tape because fibers were caught under them.

Cut from the roll a piece of tape about one inch long and hold it with the sticky side up; then, starting with the far end of the scissors blades, cut off a series of little strips across the tape that are approximately one-eighth to one-quarter inch wide, moving closer to the handles of the scissors with each cut. These small pieces of tape will stick to one blade of the scissors in an upright row. When the blade is full, simply lay the open scissors on the table and pluck off the bits of tape, one at a time, as you use them to fasten the drawings to the layout board.

First, with a couple of tabs of the tape, fasten down the edge of the drawing paper on each side of the spot where the identifying letter will be placed, to avoid a shadow cast by the edge so close to the small letter. Then the tape can be spaced as needed around the rest of the edge to hold the drawing in exact position. However, more than an inch between the bits of tape leaves the drawing vulnerable to being scraped loose or torn.

As drawings are partially anchored, remove the tabs of drafting tape to get them out of the way. But leave until last the one tab that protects the spot for the letter to be burnished into place from a sheet of "dry transfer type" (sold under such brand names as Prestype, Letraset, Tactype, and others). Futura Medium is a suitable sans-serif typeface, and "14-point" is a good size. Layouts meant to be reduced drastically should have larger letters--say, 16-point--and a correspondingly greater space allowance left below the drawings.

Where you intend to put the abc's, measure downward from the base line (or from a proximal view if one is present) and place a short dash at six millimeters with the non-photo blue pencil. On a level with the dash, set the lowercase letter and burnish it down. The final row of letters on a full-length layout will sit on the bottom boundary of the layout rectangle. On a short layout, the final row of drawings should be straight across the bottom and their letters, six millimeters below, also in a straight row. Regardless of the length of a layout, a row that is straight across the bottom is an orderly ending to a group of many different shapes. Furthermore, it is agreeably compatible with the level lines of type in the caption printed immediately below the layout.

Drawings and a scale stuck down, abc's in place, hyphens between multiple views ruled in ink; now with the drop-out blue pencil mark the layout for its repro stat. Lay the ruler outside and along the line of the left boundary. Rule a half-inch dash (like an interrupted continuation of the boundary) in the top margin of the layout board and about half an inch away from the layout area. Do the same lined up with the right boundary. Then rule a horizontal line between the vertical dashes--a line that would run from the middle of one dash to the other except that it has a gap of two or three inches approximately in the middle of the layout width. The gap need not be precisely centered. Draw arrow barbs on the ends of the line where it touches the two dashes, converting it into a pair of arrows pointing away from each other. In the gap between them, the archeologist is supposed to write explicit directions to the photostat man concerning reduction of that particular layout, such as "reduce to 6 inches," or "reduce to 37 picas." The arrows indicate the dimension that is to be reduced as directed; the two dashes that the arrow points touch mark the span of distance that the camera is to include, meaning the width of the layout.

Before you tape the keys of artifact identities to the backs of the boards, make an ordinary photocopy of each finished layout along with its key. Usually, both layout and key can be positioned on the copying machine so they will fit on the same legal-size sheet of copy paper. These copies are for the archeologist's use while he is writing captions, assigning figure numbers, and integrating illustrations as figure references in his manuscript. The small investment in copies protects the layouts from shuffling and handling that can be destructive. Indeed, the archeologist will find that paper copies are easier to work from than the mounted originals. When the copies have been made, tape each layout's key to the back of the board.

One last precaution against friction is a tissuelike piece of paper--either tracing paper or typewriter "second" sheet--to cover the drawings. Tape its edge to the top of the back surface of the mounting board--not the front--so it will hang over the front but can be turned back completely out of the way of the instructions for reduction written in the top margin.

Part 4

Varia

9 Keeping a Record of Drawings and Layouts

The prehistoric archeology that is the concern of this manual produces an astounding number of artifacts. Just one archeologist's research may involve hundreds of thousands of bits and pieces, each of which has at least a basic assigned identity and all of which must be tracked through the many steps from discovery to publication--and beyond. Depending on the sizes and complexities of the artifacts, the output of an illustrator might be five hundred or a thousand drawings a year--drawings eventually made up into many layouts. Both the drawings and the layouts thereof become that many more bits and pieces that must be tracked from discovery (or rather, creation) to publication--and beyond. Obviously, some kind of record keeping for the illustration activity should be instituted and never allowed to lapse.

If no such system exists when you, as novice illustrator, first begin work on a project, figure out a simple one for yourself and maintain it doggedly on your own initiative. It may in time be found useful by others, a backup system for their records--or lack of them. Elaboration of your initial method may become necessary as new circumstances arise; but the important thing is to make recordkeeping habitual and routine.

Illustrator's Record Book

To create your own record system, in the absence of an established one, begin by thinking over a list of information that will continue the tracking process through all the steps of your work as illustrator. Include, as well, notation categories whose data will sketch a brief log of what was done and when. Consider the following suggestions (with "examples") for your list.

The site being illustrated:

-- its complete designation ("E101B3 - Level 2")
-- its geographical name ("Wadi Kubbaniya")
-- its field season ("1978 Season")

Each artifact illustrated, the tools and the cores in separate lists:

-- the tool type ("dihedral burin") or core type ("opposed platform")
-- its personal identification ("G-14"); and alongside,
-- the number of views of it drawn

Administrative information:

-- the grant number attached to the project, if applicable
-- the archeologist in charge
-- the assistant or graduate student involved

Publication data:

-- the number of layouts made up, and the date
-- the layout measurements and the percentage of reduction intended
-- the book or journal for which they were made
-- the publication date (to be filled in later)

Like any log of continuing activity, the record should include the date on which the illustrating of that particular site subdivision was begun and, at the end of the entries of artifacts and views, the date on which the drawings were finished.

This record keeping is partly for your own benefit; the lengthening list of entries shows your progress with the immediate job. Then too, there are the small personal triumphs of saving the day in fretful situations by producing long-forgotten and now needed information. A minor, but helpful, detail: use a pencil to enter the number of penciled views of a piece; then later, as the pencil drawings are inked, go over the numeral with a pen. A glance will show how many pieces have been inked and how many are still in pencil.

A small ring binder, about six by eight inches, is convenient for this illustrator's record, in which entries are to be made every time a piece is drawn. The pages are large enough for what they must hold--short phrases and a list of letters and numerals--and the binder is small enough to lie open without crowding the drawing table. Begin each year with a new binder. Stand the old one on a shelf, like a book whose title is the year it covers; it is handy for reference and it will indeed be consulted again and again. The switch to a new binder for the binder, run off a copy of the site record in progress at year's end, so data on the whole site will be at hand as you continue the job.

When a site has been excavated, illustrated, and then, a couple of years later, excavated again, record the new material and drawings in your current log as usual, but append the year and page date of the binder where the older record of the same site can be found. In the older record, add a notation of the year and page date of your new illustration list. These date references, backward and forward in time, can be helpful in tracking the illustration history of a single site.

Ring binder pages that have been printed with a simple form are a convenience. A basic list of headings for the form might be limited to these words:

Site	Artifacts
Date	Views
Grant	Layouts
Archeologist	Published

When you have settled on the few essential headings that are most needed, it is very little trouble to make up an original of the form yourself and have it printed.

The form you design should be printed on unruled paper, for flexibility in spacing entries, rather than the more common lined notebook paper. You can buy a supply--say, five hundred sheets--of unruled filler paper meant to fit that binder, or you can investigate this alternative. A printing shop, equipped to trim and punch to order, occasionally has some paper left over from a big job; sometimes the price will be cut a little. In that case it could be cheaper to have the printing shop supply the paper than to buy the ready-made unruled notebook filler.

Ring binders usually come with a small amount of paper in them. Take a sheet of this punched paper that fits the binder, place it in the middle of a larger sheet of plain paper, and draw around it with a non-photo blue pencil. Be sure to outline the punched holes too. Within this area outlined by the blue pencil, apply the lettering from dry transfer type, such as Letraset or Prestype, to spell out the headings in whatever arrangement seems workable. For instance, "Artifacts" and "Views" will need the most space below them for their two-column list; "Date," "Grant," and "Archeologist" need little space. When it is finished, take the original of the form to the printer, along with the unruled notebook filler if you are supplying the paper.

Another necessity is a calendar of the right size to clip in the record book. Talk somebody into giving you the page from his desk calendar that has all the months of the year. On it can be marked the dates for switching from one grant or archeologist to another, deadlines for projects, and such.

That takes care of the day-to-day drawing records; however, there is another, longer range set of records to be considered, with entries made at greater intervals of time. The following suggestions on content may be helpful.

Master List of Sites Illustrated

This kind of list becomes a necessity when sites and their illustrations multiply. Begin the list as a rough draft--informal, handwritten, with notations freely inserted as needed. This will allow the format and the type of information most useful to your operation to evolve naturally. The important thing is to keep the information up to date. Perhaps once or twice a year, the list should be typed (or retyped), with the handwritten notations on changes of status incorporated. If the illustrator serves several archeologists, a separate record of this kind should be set up for each.

This is a continuous list on which to enter pertinent information as each site subdivision passes through the steps from drawing to publication. It is both inventory and status sheet--and becomes history as it grows longer. Include such data as the following:

- -- The complete site number.
- -- The year of the field season that produced the material.
- -- The local name of the site.
- -- The number of layouts of tools and of cores.
- -- The month and year the layouts were mounted.
- -- If the drawings have not been mounted, note that fact and the date the drawings were finished.
- -- If the original drawings or layouts are no longer available--perhaps sent out of the country for publication--note where they were sent and the date.
- -- When layouts have been published, enter the name of the book or periodical, and the date.

An alternative, but more complex, plan is setting up a card file. It may be a desirable addition if time and help are no problem; however, the list method is quick and easy, and its very informality tends to ensure that entries are kept up to date. Grab it to add an item, grab it to check something; it is simpler.

Another category of information, especially useful to an illustrator, is the measurements used for the layouts that were made up for various books and journals and the amount of reduction for which they were intended. List these reminders alone and keep them on file. This record can save time at a future date when a new book with the same format is in preparation or an article is being readied for submission to one of the same publications.

If ever the time comes when there are no artifacts to draw, no layouts to make up, no records to fill out or shelves to tidy, here is a suggestion for a small, tedious research project--another list. This one would assemble, for easy reference, the practical information about the handful of scientific periodicals to which the illustrator's archeologists most often submit their articles. It calls for some library time spent examining copies of the journals, noting their formats, and measuring in each one the effective page's width and length--column width, too, if pages are so divided. Other characteristics to observe are mentioned in the chapter on layout composition.

10 The Working Relationship of Archeologist and Illustrator

This is a plea to the archeologist on behalf of the illustrator, and vice versa. Assuming you both seek a superior final product, a combination of understanding, planning, and cooperation is more than merely desirable if this joint effort is to move smoothly--and preparing for publication is indeed a joint effort. To this end, be aware of the problems of each other's specialty.

Illustrator, an impulsive rush to the archeologist with each one of your indecisions at the moment it arises is thoughtlessly intrusive and disruptive. Save up the small questions about the artifacts. Common examples are the orientation of a strange piece, the direction of blow for a flake scar, the number of arrows for a burin with many facets. If you group problem drawings and artifacts, each with a slip noting the question, it will save time in getting the answers. Try to work around the problems until you can meet with the archeologist. Leave him a note saying you have accumulated some questions and ask for two or three minutes of his time to deal with them. The intent is to keep from interrupting his work again and again; but it is also an attempt to get a few uninterrupted moments of his attention focused on answering illustration questions.

Be considerate in giving the archeologist some "lead time." When you see that you will soon be ready for more work, notify him a day or two in advance; he must find the time in his schedule to select more artifacts for illustration. Meanwhile, you can keep drawing while he determines what else he needs to have drawn.

If you happen on a piece that seems to have been misclassified or tossed into the wrong box, call it to the archeologist's attention and state what you believe is evidence of a mistake. If he confirms the original classification, accept it. You have an obligation to catch possible accidental errors and report them, but what the archeologist says goes. Never argue.

When your illustrating services are scheduled to several archeologists in rotation, you will become aware that they differ in their working methods and approaches to typology. A certain flexibility in adjusting to this variety of judgments and leanings must be developed.

Even after years of experience, it is still a guess how long it will take to draw a particular batch of artifacts. At least four major factors weigh heavily in the estimating: the number of views each piece must have, the size range of the pieces, the complexity of the retouch, and the amount of cortex area. As a rough rule of thumb, the more secondary views, the more big pieces, the more flake scars, and the more cortex to be stippled, the longer the drawing time necessary for a group of assorted artifacts. Archeologists should be aware of these considerations when gauging both the budget and the amount of material they want illustrated. Since the artist can be only marginally helpful in this, it is a prudent plan for the archeologist to group the pieces as to priority--a first priority for the "must haves," a second for the "all possibles," and a third for any time left. It will ensure at least minimum, and probably adequate, representation across the collection. This setting of priorities is especially advisable when a very limited budget can cover only three or four weeks of the artist's time. If the archeologist also wants layouts of the drawings made up for publication, it must be a part of the plan from the beginning, and time for it must be allotted toward the end of the job span. When the illustrator and the archeologist can discuss the general outlines of a project, with mutual understanding of the limitations imposed by specific factors, the subsequent informed and purposeful efforts by the artist work to the benefit of the archeologist.

When you, the illustrator, have become experienced in drawing and knowledgeable enough about tool types not to need constant supervision, a working arrangement with more responsibility is possible--one that fits well with the awkward schedules of archeologists who must spend months in the field. For those times when no archeologist will be available for consultation and yet a collection of artifacts must be drawn, it is a good plan for the illustrator to have access to the field notes involved. The classifications, measurements, and descriptions of the pieces form an authoritative guide to the archeologist's judgments on significant attributes of the artifacts--especially important if it is his method to leave mixed bags of artifacts "for illustration" straight from the field, or a box full of selected nameless tools bearing only catalog numbers. An occasional tool can be ambiguous to an artist; one might make a case for either of two classifications except for the firm opinion of the archaeologist in charge. Eventual consultation, after his return, will go more smoothly if the illustrator will tag any problem pieces on which there are still questions and insert temporary markers at their descriptions in the field notes.

A final word in the form of a reminder to the archeologist: please don't leave town without checking to see that your illustrator is well supplied with artifacts to draw while you are gone.

Appendix 1
Specialized Terminology

This limited list provides a minimal background of essential terms peculiar to lithic technology, tool typology, and lithic illustration; a few of the expanded definitions tie together all three. Additional explanations are interwoven with the manual's instructions. Since some knowledge of these terms is needed for understanding the text, the definitions are broadly simplified to serve illustrators rather than archeologists. More precise and exhaustive discussions of the scientific terms can be found in the glossaries of archeology books, some of which are listed in Suggested References.

Artifact. Anything made by man. The term is used here to mean any piece of stone that has been modified by man.

Axis of detachment. The path of the force that removed the piece from the core. It runs from the point of impact on the platform of the artifact, toward the distal end.

Backing. Very steep retouch along a lateral edge, forming a near ninety-degree angle with the dorsal face of the artifact. The retouch is usually obverse, but it may be inverse or a mixture of the two. It has the effect of thickening, blunting, and shaping the lateral edge and is somewhat analogous to the blunt back edge of a table knife. Backing may straighten, curve, or shoulder the outline shape of an artifact. Long, narrow perforators are often "double backed" (meaning both lateral edges) to that shape, producing a strengthened point.

Blade. An elongated piece of stone that was detached from a core and is, by oversimplified definition, at least twice as long as it is wide. When detached from the blade core, the removals, one after another, usually produce on the core a series of flake scar ridges, approximately parallel; they in turn become dorsal surface features trending lengthwise of the blades that are subsequently detached. However, the pattern of flake scars on a blade is not limited to lengthwise configuration.

Break. Evidence of accidental damage that resulted in the loss of an appreciable part of an artifact and altered its length, width, or shape. In most cases it is the distal or the proximal part that is missing.

Bulbar surface. Another term for the entire ventral surface of an artifact. "Bulbar" refers to the bulb of percussion, a ventral feature, but its meaning extends to the rest of the surface, as well; the bulb itself and the platform of the piece may be missing, removed by retouch, a snap, a burin blow, or simply a break.

Bulb of percussion. The swollen area on a ventral surface where it meets the striking platform. The meeting of the two is the spot that received the force (either a blow or pressure) detaching the piece from the core. Ranging in prominence from barely perceptible to rotund, the bulb of percussion is the primary feature that identifies the ventral surface of a flake or blade. Locating its position reveals which is the proximal end of an artifact. Other determinations about tool type and correct orientation for drawing the piece are dependent on relationships to "proximal" and "distal."

Burin facet. The scar formed by the detachment of a burin spall.

Burin spall. The piece struck off to produce a burin. An occasional burin spall may show subsequent modification, by retouch, into a perforator.

Carinated. Shaped like the keel of a ship.

Cleavage surface. A surface formed by a separation of the rock along a natural cleavage plane or crack. Among the raw materials most apt to exhibit cleavage surfaces are quartz crystals, slate, and some kinds of petrified wood.

Core. A pebble, cobble, slab, or chunk of rock from which pieces were removed, by percussion or pressure, to be made into tools.

Cortex. The natural, weathered, exterior surface of the rock, as opposed to an interior or flaked surface.

Cross section. The area that would be exposed if an artifact were cut in two. For a "transverse section," the imaginary cut is made parallel to the base line of the artifact drawing; the outline of the section keeps at its top the front surface of the view that has been sectioned. For a "lateral section," the cut is made perpendicular to the base line of the artifact drawing; the outline of the section is oriented and placed like a profile view, but in horizontal alignment with the points through which the cut was made.

Debitage. The waste material that is the by-product of the manufacturing process--the pieces not turned into tools by further modification. An example of debitage is the flakes of stone removed in the preparation of cores. Small chunks and chips, as well as the waste material from the shaping of tools by retouch and the sharpening or reworking of damaged tools, are termed "debris."

Denticulated. "Toothed" by retouch that forms a connecting series of three or more intersecting notches along the perimeter of a piece.

Direction lines. The nested series of curved lines drawn in each flake scar to indicate, by their placement, the direction of the blow or pressure that produced that scar when the force detached a piece of the rock.

Distal end. The extremity of the artifact opposite the proximal end or bulb of percussion. It is identified by examining the ventral surface.

Dorsal surface. The surface of an artifact that exhibits either cortex or flake scars resulting from removal of material before detachment of the piece from the core. It is the opposite of ventral. In illustrating, the dorsal surface is usually the first drawn--variously termed the "primary," or the "principal," or the "front" view.

Flake. A piece that has been detached from a core and is less than twice as long as it is wide. "Flake" is also used in a general sense to mean any fragment of stone detached by percussion or pressure and exhibiting a bulbar surface.

Flake scars. The marks left on the surface of an artifact by the removal of flakes of the rock with intentional blows or pressure. Their boundaries are ridgelike and range from sharply prominent to scarcely more than a change in rock texture. Flake scars are normally produced in succession and overlap, truncate, or encroach upon neighboring scars.

Geometrics. A general category of artifacts that includes lunates (crescent-shaped), triangles (three sides), trapezes (four sides, two approximately parallel), and rectangles (four sides). They are very small tools, usually less than an inch long. Their distinguishing geometric shapes are formed by backing and a sharp cutting edge that is approximately straight. Draw all geometrics with the cutting edge perpendicular to the base line. The platforms and bulbs will be missing, but orient the pieces so that the bulbar end is toward the bottom of the drawing. (Exception: *see* Scalene triangle.)

Halfa. The name attached to a type of prepared core and to the final piece struck from it, whether or not the piece is retouched into an obvious tool. The preparatory work is like that of a Levallois core with the exception of a diagnostic series of parallel bladelet removals from the distal end, on the face of the core. The ultimate desired piece, struck from the proximal end of the Halfa core, displays remnants of these parallel bladelet scars at its distal end. A butt view of the piece is customary.

Handaxes. More commonly called "bifaces." These large tools are rather flat, with outlines that narrow toward the distal end if they are not, indeed, pointed. Among the many subtle variations of outline that have descriptive names are "amygdaloid" (almond-shaped) and "cordiform" (heart-shaped). The flake scars that shaped them spread over most or all of both of the major opposing surfaces; hence "bifaces" because of their "bifacial" retouch.

Hinge fracture. The abrupt and steeply prominent termination of the distal end of a flake scar, an effect that resulted when the force of detachment encountered increased resistance in the rock or changed direction.

Krukowski microburin. Backed like a La Mouillah point, but the microburin scar runs back onto the body of the bladelet.

La Mouillah point. Like a *piquant-trièdre* except it was backed and then the tip was twisted off in such a way that the microburin scar forms an extended point.

Levallois. The name attached to a type of prepared core and to the final piece struck from it, whether or not the piece is retouched into an obvious tool. The preparatory work on all parts of a Levallois core follows certain patterns of removals to shape it before the striking off of the desired piece for a tool. The scar remnants of these preparatory removals, visible on the dorsal surface and the butt of a Levallois piece, are usually diagnostic of its production by the Levallois technique. A view of the platform of a Levallois flake or blade is expected. (Approximate pronunciation: "Luh-val-wah.")

Lunate. A geometric microlith, somewhat crescent-shaped. The curved part of its outline is formed by backing and is therefore blunt and thickened; the sharp cutting edge is approximately straight, sometimes slightly concave or convex.

Microburin, microburin technique. The microburin is not a tool but a waste product of the "microburin technique" of shaping such tools as lunates, triangles, and trapezes. A bladelet is notched; a piece (the microburin) is twisted off at the notch and discarded; then the remaining portion is further modified to the desired shape by retouch or backing. In another version of the microburin technique, a bladelet is first backed along one edge, then the twisting-off action is performed at the backing instead of at a notch. (*see* La Mouillah point, *Piquant-trièdre*, and Krukowski microburin.)

Microliths. Very small stone tools, usually less than an inch long.

Oblique. At an angle to both the vertical and the horizontal.

Order of blows. Evidence of the sequence in which flakes of stone were removed and therefore information related to the artifact's pattern of manufacture. As each successive flake is removed from a piece of rock, whether by intentional blow or pressure, its scar encroaches upon adjacent scars and obliterates portions of the earlier ones.

Orientation. How an artifact is positioned for illustration, according to its type and attributes. Specified orientation of tools and cores in general, and certain ones in particular, is a part of the code and conventions of lithic illustration.

Patination. The visible changes wrought by time and exposure on the flaked surfaces of rock. It appears as an alteration in the color, luster, and texture. A particular flaked area on an artifact, when compared with other flaking on the same piece, may exhibit marked contrast in degree of patination, suggesting that the scars differ emphatically in age. In illustrating such a piece, the greater patination of the "older" work is indicated by the marks specified for drawing heavily patinated scars, distinguishing them from "younger" work.

Piquant-trièdre. A bladelet that has been notched and then had a microburin twisted off at the notch. The microburin scar forms a thin extended point for the *piquant-trièdre*. The fragment that was twisted off and discarded (the microburin) has a microburin scar that spreads back onto the surface of the fragment.

Platform, striking platform. The area on a core that received the blows or pressure that removed pieces for tools; also, the remnant of that surface at the proximal ends of the pieces that were detached. Cores may have one or several platforms and are classified, in part, according to the number, type, and location of their platforms.

Projectile points. A general classification that includes dart points, arrow points, and spear points.

Proximal end. The extremity of the artifact that has, or once had, the bulb of percussion. It is identified by examining the ventral surface.

Retouch. All intentional flaking that modified a piece after detachment from a core. Most commonly, retouch is a series of contiguous flake scars, usually small, that originated at the perimeter of the artifact and served to sharpen, blunt, thin, shape, or otherwise modify the piece. They may lie flatly on the surface, meet it steeply, or scoop deeply into it. A repetition of retouch along the same edge removes portions of preceding retouch scars. Flake scars directed toward the dorsal surface, by force applied to the ventral surface of the artifact's edge, are *obverse retouch*. Flake scars directed toward the ventral surface, by force applied to the dorsal surface of an artifact's edge, are *inverse retouch*.

Scalene triangle. A variety of triangle with three unequal sides. The longest side is the sharp cutting edge, and the other two sides are formed by backing. Draw scalene triangles with the longest point at the top of the drawing and the sharp edge perpendicular to the base line, regardless of bulb position.

Scar ridges. The boundaries of individual flake scars.

Selective grid. A network of horizontal and vertical guide lines, ruled lightly in pencil, their spacing tailored to the individual artifact's shape and features, to aid in drawing successive multiple views of the same piece.

Serrated. Retouched along a cutting edge to form very small, saw-like teeth.

Sickle sheen. The glassy brilliance of certain areas on some few artifacts, caused by the silica present in vegetable matter such as grasses. Distribution is usually along part of a cutting edge of a blade or bladelet. Sickle sheen is not to be confused with polish caused by wear. The latter may be smooth and glossy, but sickle sheen reflects light with a distinctive dazzling glare.

Snap. A purposeful break that was a step in the manufacture of a tool. For example, in the process of making a "burin on a snap," the surface of the snap was used as a striking platform for the burin blow.

Sur enclume (bipolar, bidirectional) backing. Backing produced by force from two directions. In contrast, the more common obverse ("normal") backing and inverse backing are produced by force applied from a single direction.

Tools. The term as used here includes weapons (projectile points). Although such by-products of manufacture as microburins and burin spalls are not considered "tools," they are illustrated and therefore are discussed in this manual.

Trapeze. A geometric microlith whose outline has four sides: the two lateral sides are nearly parallel, the longer one being the sharp cutting edge; the shorter, opposite edge may or may not be backed. The remaining distal and proximal sides are backed obliquely, at an angle both to the cutting edge and to each other.

Triangle. A geometric microlith, triangular in outline. One side of the triangle is the sharp cutting edge; the other two sides are shaped by backing. The backed sides may be straight, concave, or convex.

Truncation. The process that produced, and the tool type that exhibits, drastic redirection of outline in the artifact's distal or proximal area and was effected by steep retouch. A truncation may be straight, convex, or concave; its location distal, proximal, or oblique to either; the retouch that produced it, obverse or inverse.

Tucked view. A type of secondary partial view that appears to be "tucked" behind the primary view of the artifact. It is a space-saving way to show features near the edges of ventral surfaces.

Use retouch. An irregular scattering of small scars often found on the edges of artifacts and thought to be the accidental result of tool use. The flaking was not purposeful and therefore is not considered "retouch." Use retouch is omitted from the drawings, as a general rule, except in the case of pronounced concentrations of it that might indicate prolonged use of a particular portion of an edge; then, careful copying of the characteristic irregularities will distinguish the flaking from purposeful retouch. Another occurrence that might be classed as "use retouch," but would be included in the drawing, is the scar of an impact spall, sometimes found on the distal end of projectile points.

Ventral surface. The surface of the artifact that was created by the separation of the piece from the core and, as a result, has (or once had) a bulb of percussion. It is the opposite of dorsal. In illustrating, the ventral side of an artifact is often loosely termed "the back" of the piece or "the back view." Another term for it is "bulbar surface."

Vertical axis. For illustrating purposes, an imaginary line that is perpendicular to the base line and runs through the center of the artifact as it is oriented for drawing. It is the axis upon which the artifact is rotated--a one-quarter turn for a profile, a one-half turn for a back view--so that each feature and flake scar maintains its distance above the base line in views of the front, profile, and back.

Weathering. Alteration of the artifact's surface by natural agencies, such as moisture, heat, cold, and wind-driven particles. In extreme examples, all projections and edges of an artifact may be rounded off and flake scars blurred almost beyond identification.

Appendix 2
The Necessary Art Materials and Tools

Table. Any steady, smooth table of comfortable height and with sufficient space; a flat-topped drafting table is ideal.

Stool. Adjustable in height.

Light. General illumination.

High-intensity lamp. Absolutely essential; do not get fluorescent; a good quality two-speed (Tensor, for example), run on low, will outlast several cheap one-speeds that tend to overheat and fuse to their bulbs; keep a supply of bulbs.

Magnifier. Essential; four-inch on a weighted base is ideal; if necessary, rig one with a "reading glass."

Reducing glass. Desirable frill but not essential.

Pencil sharpener. One that sharpens with a long taper; an electric sharpener if possible.

Paper cutter. Long, strong blade, the best affordable; at least put it on your wish list.

Vaporizer. Sold in pharmacies, meant for infants; to increase humidity while inking if air is extremely dry.

Drawing paper. Strathmore single-ply, medium surface; sold in large single sheets; use for all artifact drawings.

Drawing pencils. The 5H for all artifact drawings except when dummy layout procedure is to be followed; then use a 2H so photocopy machines will pick it up.

Non-photo blue pencil. Also called "drop-out blue"; not the same thing as an ordinary blue pencil; used for ruling the layout rectangle because the camera is blind to it.

Eraser. Pelikan brand plastic eraser; white, translucent, easy on the paper's surface; some imitations stick.

Erasure template. Helpful for errors amid complex retouch.

Pen for drawing. Hunt's no. 104 penpoint and penstaff for inking all artifact drawings; these simple little penpoints are inexpensive, so buy six to a dozen at a time, but select personally by feeling each one and rejecting those with splayed or twisted tips and roughness that will snag fibers from penwipers.

Pen for ruling. A reservoir type of drafting pen (for example, Leroy or Rapidograph), with a double-zero point, for ruling hyphens between artifact views, ticks to indicate cross sections, and for scales; not for drawing artifacts.

Pens for writing. Suggest a black Pilot "Razor Point" for records and labels; add a red one to make labels emphatic.

Ink. Pelikan brand standard, waterproof black india ink for all drawings (not the kind developed especially for use in outer space; it is too fluid for the little pen).

Penwiper. Multilayer, lint-free cocktail size paper napkins.

Pen cleaner solution and squeeze bulb. K&E brand; the kind like liquid soap; for soaking reservoir pens and the 104 penpoints.

Plastic rulers. One-foot, clear plastic, marked with inches on one edge, centimeters on the other, and an extra scale down the middle with zero centered; good quality with the printing laminated inside the plastic--not on the surface (this is important); two such rulers to use in concert are convenient.

Wooden ruler with metal edge. A twenty-four inch is good; needed when showcard board for layouts must be cut with a knife; ideal is the kind meant for printers, with a pica scale as well as inches (useful for measuring and marking art work when picas, instead of inches, are the units of measure specified).

Triangles. Two clear plastic right-angle triangles; one about six or eight inches, one about fourteen inches; beveling is immaterial because a row of drafting tape patches will lift a flush edge for use with a reservoir pen.

Proportion scale. Valuable for figuring layout proportions, measurements, and reduction; buy at art supply stores; mathematics deficiency overcome for a few dollars.

Scissors. One pair, any size, for cutting pieces of drawing paper and general use; another pair, small, with slender blades, for layout mounting; essential.

Knife. X-Acto knife with pointed, triangular blade; keep a packet of spare blades; for scraping out ink blobs and remaking old layouts.

Carpet knife. Inexpensive kind with razorlike blade and plastic grip; for cutting big sheets of showcard board that are too large for the ordinary paper cutter; use in conjunction with a metal-edged ruler.

Tweezers. The type with a slanted, squared-off business end; cheap eyebrow tweezers are fine.

Burnisher. Pencil-size friction and pressure tool for transferring lettering from sheets of Letraset, Prestype, and so on; fine orangewood ones are extinct, plastic ones are current.

Typewriter brush. The kind like a wooden pencil, eraser on one end, stubby brush on the other; for cleaning grit and lumps of calcium carbonate from artifacts.

Drafting brush. Short-handled; for whisking away eraser crumbs.

Drafting tape. Scotch brand, half-inch width much preferred; this is not "masking tape"--they look alike but perform quite differently and are not interchangeable.

Scotch brand Magic Transparent Tape. The frosted kind--not the shiny; half-inch width if possible; enclosed, hand-held, plastic dispenser ideal--not the weighted table model; for mounting the drawings on the layout boards.

Rubber cement. A small bottle for making rubber cement pellets to prop artifacts in position; do not use for mounting drawings.

Fingernail polish and remover. Clear, colorless liquid fingernail polish to protect the identification that is inked on artifacts; remover to thin the polish and to clean off the excess from careless applications; bright tomato-red polish for a dot to mark artifacts "illustrated" when returned to collections.

Rigid board for layouts. Strongly suggest 14-ply hot press (slick) showcard board; sheets 28" x 44". This is the material on which arrangements of drawings are to be mounted for publication; flexible "poster board" is too springy.

Dry transfer type. Ready-made lettering for the abc's on layouts (Letraset, Prestype, or others); 14-point is a good type size; Futura Medium is a suitable, widely available sans serif typeface (a plain style without twiddlybits).

Other Supplies

Ring binder. For keeping illustration records; check chapter 9, on recordkeeping, for details of binder and record forms.

Ordinary "yellow paper" or equivalent. Standard size, on which to tape drawings temporarily for filing.

File folders. To hold sheets of drawings.

Brown envelopes. To store the finished, mounted layouts of each site subdivision; clasps are unnecessary and will break off almost immediately anyway.

Ball of twine (unless you save string). For loosely tying groups of brown envelopes of layouts, to segregate them within stacks by total site numbers and field season; the twine is a simple retrieval aid; layouts must be stored flat.

Scratch pads. Suggest package of a dozen, 4" x 6"; for labels, tags, memos, and such.

Flat sacks. All sizes, saved from shopping; for storing negatives, extra stat prints, and such.

Corrugated board. Salvaged from cartons and boxes; to be used for stiffeners.

Paper clips, rubber bands, painter's plastic drop cloths, plastic bags, bookshelves, antiseptic, Band-Aids, and a sewing kit (strands of thread taped on will divide up a complicated hammerstone or potsherd).

Now you are in business; this is everything you need except determination, tenacity, and patience.

Appendix 3

Useful Trivia

Oak tag. A smooth-surfaced, semirigid paper called "oak tag" is an ideal disposable covering for the immediate working area on a drawing table. It comes in sheets about the size of a large desk blotter. In addition to protecting the tabletop from scratches and ink stains, it has a pleasant subdued color, somewhat like manila envelopes, that decreases the glare of the lighted surface the illustrator must face while drawing.

Drop cloths. Clear plastic drop cloths, meant to keep paint spatters off furniture, are inexpensive protection for artwork. The 9' x 12' size gives generous cover, between work sessions, for a table spread with drawings or layouts. The waterproof aspect is insurance against roof leaks and plumbing disasters.

Carpet knife. An inexpensive version of a carpet knife with a plastic grip, meant for do-it-yourself carpet laying, is a passable substitute when no paper cutter large enough to take the big sheets of showcard board is available. The carpet knife's plastic grip is more comfortable to hold and easier to control than a pocketknife. Guide its cut with a metal-edged ruler.

Typewriter brush. Every lithic illustrator's personal tool kit should contain the kind of typewriter eraser that is encased in wood, like a pencil, and has a small stiff brush on one end. Both ends are useful for cleaning from artifacts any residue of dirt or calcium carbonate caught in the retouch. Push off small lumps with the eraser and whisk away grit with the brush.

Furniture damage. Beware of laying a plastic eraser on certain furniture finishes. Even overnight contact can cause some types of varnish, lacquer and plastics to soften and become like syrup.

Eraser storage. When several plastic erasers are bought at once, wrap each separately in aluminum foil, then roll it tightly in a plastic sandwich bag to prevent drying out and hardening over a period of months.

Tape tabs. When tabs of drafting tape are used to fasten the flap of a storage envelope that will be opened from time to time, it is a thoughtful refinement to turn under a bit of the tape's end to make it easier to grasp for unsticking.

Clear fingernail polish. If the catalog identification on an artifact has not been covered with a protective coating, it is almost certain to smear or rub off during the prolonged handling incident to the drawing process. Clear nail polish, too thick in its original form, can be thinned about one-third with either acetone or nonoily nail polish remover.

Hidden data. Be sparing and careful in applying the protection of clear fingernail polish to catalog numbers on artifacts. Often it is the illustrator's intent examination that discovers signs of use (such as use striations and sickle sheen) overlooked by the archeologist. This evidence will be missed by both if likely areas of surface are covered by the coating of polish.

Tweezers. Ordinary eyebrow tweezers are an efficient tool for wielding a miniscule twist of tissue wet with remover, when careless application of clear fingernail polish over the site number on an artifact has made flake scar ridges too slick to accept pencil marking.

Red fingernail polish. As soon as artifacts have been drawn, and before they are returned to the main body of the collection to be mixed with unillustrated pieces, apply a dot of bright tomato-red fingernail polish to each. This eye-catching mark tags these artifacts as "drawn" when additional specimens are sought later for illustration. Also, retrieval is quicker when a researcher wants to examine certain pieces whose published drawings have prompted interest.

Color code. For those occasions when items on display will be handled--for instance, a conference of visiting archeologists--color coding the artifacts by site will make it easier to spot any pieces wrongly replaced and faster to return specimens to storage when participants have departed. Fingernail polish is a quick marker, stays on during handling, and can be removed later if desired. In this case use dot combinations of some of the unusual colors available--yellow, orange, green, purple, blue--but stay away from bright red, your standard dot color that means "drawn."

Obsidian problem. Pronounced ripples and vitreous dazzle under direct light can make some obsidian artifacts as hard to "read" as a pile of black spangles. Individual flake scars must be puzzled out and marked. Neither drawing pencil lead nor ordinary colored pencils will stick to the glassy surface. Try a red or orange china-marking pencil--the soft, waxy kind used to write prices on tableware.

Edge marking. Pencil lead marked on the edge of an artifact that is made of very light colored raw material will help show up the exact edge characteristics against the glare of the drawing paper.

Ink bubbles. Drawing ink, undisturbed for some hours or days, may not seem as intensely black as when it was agitated by use. Instead of shaking it--which fills it with bubbles that pop specks of ink from the pen to the drawing paper--hold the bottle in contact with the table top while making tight little quarter-inch circles for a minute to stir the ink to full blackness without producing bubbles.

Penpoint in use. If a group of penciled drawings from a single site are being inked when work on that job must be discontinued for weeks, or even months, wrap the penpoint you are using and store it with the artifacts themselves. When inking of the penciled drawings is resumed, continuing the work with the original penpoint will ensure a better match with the earlier work than inking with a different point that would not have the identical degree of wear.

Eye level. For help in discovering the points at which a profile view should touch the vertical guide line, lay the artifact on a portable flat surface (your plastic triangle is ideal) and lift it to eye level. The points that touch the surface and the parts that may arch free of it are immediately apparent.

Cobble-shaped pieces. It is difficult to draw the outline of large spheroid pieces like hammerstones and thick cores; the least head movement while sighting straight down on them compounds errors. Stand your right-angle triangle on its bottom edge and back its vertical edge up to the artifact as it rests on the drawing paper. Make a little mark on the paper at the corner of the triangle. Again, back the standing triangle up to touch the artifact a bit farther along, and mark at the triangle's corner as before. Repeat this at close intervals all around the perimeter of the artifact. You will then have the size of the piece indicated on the paper by a series of small dashes. After that, by sighting straight down over one portion of the perimeter at a time and adjusting your line of sight to make that portion fit the dashes, you can refine them into a true outline.

Ruling with ink. When ruling a line with a reservoir drafting pen, the ink can catch and spread under a flat, unbeveled ruler whose edge lies in contact with the paper. A series of patches of drafting tape, two layers thick, stuck on the underside of the ruler one-eighth inch from the edge will raise a flat ruler just enough to be used with a pen. The same arrangement of patches can be applied under, and back from, the edge of a triangle that is to function as a straightedge with a ruling pen. Patches are better than a continuous strip of tape because they obscure less of the artwork immediately adjacent to the line and otherwise visible through the transparent plastic.

Suggested References

Readings and Technical References

Bordes, F. 1961. *Typologie du paléolithique ancien et moyen*, 2d ed. Publication de l'Institut de Préhistoire de l'Université de Bordeaux, Mémoire no. 1. Bordeaux: Delmas. [Illustrated in part by Pierre Laurent. An authoritative reference for typology and the orientation of artifacts for illustration, especially bifaces.]

__________. 1970. *The Old Stone Age*. New York: McGraw-Hill. [Illustrated by Pierre Laurent. Although it is a convenient small paperback and is written in English, it does not entirely replace Bordes 1961, because the illustrations are small.]

Brezillon, M. 1968. *La dénomination des objets de pierre taillée: matériaux pour un vocabulaire des préhistoriens de langue française*. Gallia Préhistoire, suppl. 4. Paris: Centre National de la Recherche Scientifique.

Cahen, D., L. H. Keeley, and F. L. Van Noten. 1979. Stone tools, toolkits, and human behavior in prehistory. *Current Anthropology* 20:661-83.

Ceram, C. W. 1951. *Gods, graves, and scholars: The story of archaeology*. New York: Alfred A. Knopf. [Also available in paperback.]

Clark, J. D. 1954. *The prehistoric cultures of the Horn of Africa*. Cambridge: Cambridge University Press.

Crabtree, D. E. 1972. *An introduction to flintworking*. Occasional Papers, no. 28. Pocatello: Idaho State Museum.

Daniel, G. 1976. *A hundred and fifty years of archaeology*. Cambridge: Harvard University Press.

Dauvois, M. 1976. *Précis de dessin dynamique et structural des industries lithiques préhistoriques.* Périgueux: Pierre Fanlac. [Exhaustive, esoteric depictions of flintknapping details. Elaboration beyond that necessary for a novice illustrator.]

Jennings, J. D. 1968. *Prehistory of North America.* New York: McGraw-Hill.

Keeley, L. H. 1980. *Experimental determination of stone tool uses: A microwear analysis.* Chicago: University of Chicago Press.

Laplace-Jauretche, C. 1957. Typologie analytique: Application d'une nouvelle méthode d'étude des formes et des structures aux industries à lames et lamelles. *Quaternaria* 4:133-64.

Laurent, P. 1970. Les sections et les coupes dans le dessin d'archéologie préhistorique. *Origini* (Rome) 4:7-44. [Valuable article on drawing all types of cross sections. For other illustrations by Laurent, see Bordes 1961 and 1970.]

MacNeish, R. S., A. Nelken-Terner, and I. W. Johnson. 1967. *The prehistory of the Tehuacan Valley. Vol. 2: Nonceramic artifacts.* Austin: University of Texas Press.

Marks, A. E., ed. 1976. *Prehistory and paleoenvironments in the Central Negev, Israel. Volume I: The Avdat/Aqev area, part 1.* Institute for the Study of Earth and Man, Reports of Investigations, 2. Dallas: Southern Methodist University Press.

Montet-White, Anta. 1968. *The lithic industries of the Illinois Valley in the Early and Middle Woodland period.* Anthropological Papers, Museum of Anthropology, University of Michigan, no. 35. Ann Arbor: University of Michigan.

Movius, H. L., Jr., N. C. David, H. M. Bricker, and R. B. Clay. 1968. *The analysis of certain major classes of Upper Palaeolithic tools.* Bulletin no. 26. Cambridge: American School of Prehistoric Research, Peabody Museum, Harvard University. [Gives needed background and supplies much information for the novice illustrator who cannot read French. Excellent drawings of specific tool types, diagrams of their manufacture and of their attributes.]

Odell, G. H., and F. Odell-Vereecken. 1980. Verifying the reliability of lithic use-wear assessments by "blind-tests": The low-power approach. *Journal of Field Archaeology* 7:87-120.

Piggott, S. 1979. *Antiquity depicted: Aspects of archaeological illustration*. London: Thames and Hudson.

Ridgway, J. L. 1938. *Scientific illustration*. Stanford: Stanford University Press.

de Sonneville-Bordes, D., and J. Perrot. 1954. Lexique typologie du paléolithique supérieur: Outillage lithique--I, grattoirs--II, outils solutréens. *Bulletin de la Société Préhistorique Française* 51:327-35.

_________. 1955. Lexique typologie du paléolithique superieur: Outillage composite--III, percoirs. *Bulletin de la Société Préhistorique Française* 52:76-79.

_________. 1956a. Lexique typologie du paléolithique supérieur: Outillage lithique--IV, burins. *Bulletin de la Société Préhistorique Française* 53:408-12.

_________. 1956b. Lexique typologie de paléolithique superieur: Outillage lithique--V, outillage à bord abattu--VI, pièces tronquées--VII, lames retouchées--VIII, pièces variés--IX, outillage lamellaire, pointe azilienne. *Bulletin de la Société Préhistorique Française* 53:547-59.

Suhm, D. A., and E. B. Jelks, eds. 1962. *Handbook of Texas archaeology: Type descriptions*. Published jointly as Texas Archaeological Society, Special Publication no. 1 and Texas Memorial Museum, Bulletin no. 4. Austin, Texas: Texas Archaeological Society and Texas Memorial Museum.

Tixier, J. 1963. *Typologie de l'Epipaléolithique du Maghreb*. Mémoires du Centre de Recherches Anthropologiques, Préhistoriques et Ethnographiques, 2. Paris: Arts et Métiers Graphiques. [An authoritative reference for tool typology and the accepted orientation of many artifact types.]

________. 1974. *Glossary for the description of stone tools, with special reference to the Epipalaeolithic of the Maghreb.* Newsletter of Lithic Technology, Special Publication, 1. Pullman: Washington State University.

Tixier, J., M.-L. Inizon, H. Roche, and M. Dauvois. 1980. Préhistoire de la pierre taillée I: Terminologie et technologie. Valbonne: Cercle de Recherche et d'Etudes Préhistoriques. [Archeological terms in both French and English.]

White, Anta M., L. R. Binford, and M. L. Papworth. 1963. *Miscellaneous studies in typology and classification.* Anthropological Papers, Museum of Anthropology, University of Michigan, no. 19. Ann Arbor: University of Michigan.

Willey, G. R. 1966. *An introduction to American archaeology. Vol. 1: North and Middle America.* Englewood Cliffs, N.J.: Prentice-Hall.

________. 1978. *Excavations at Seibal, Department of Petén, Guatemala: Artifacts.* Memoirs, vol. 14, no. 1. Cambridge: Peabody Museum of Archaeology and Ethnology, Harvard University.

Zweifel, F. W. 1961. *A handbook of biological illustration.* Chicago: University of Chicago Press. [Some production information applicable to archeological illustrating.]

Sources of Additional Drawings

The following is an abridged list of works containing hundreds of my lithic artifact illustrations, including examples of additional tool and core types.

Close, A. E., F. Wendorf, and R. Schild. 1979. *The Afian: A study of stylistic variation in a Nilotic industry.* Dallas: Department of Anthropology, Institute for the Study of Earth and Man, Southern Methodist University.

Henry, D. O. 1976. The excavation of Hayonim Terrace: An interim report. (with contributions by Arlette Leroi-Gourhan). *Journal of Field Archaeology* 3:391-406.

_________. 1978a. *The prehistory and paleoenvironment of Hominy Creek Valley, 1977 field season.* Tulsa: Laboratory of Archaeology, Department of Anthropology, University of Tulsa.

_________. 1978b. Big Hawk Shelter in northeastern Oklahoma: Environmental, economic and cultural changes. *Journal of Field Archaeology* 5:269-87.

_________. 1982. The prehistory of southern Jordan and relationships with the Levant. *Journal of Field Archaeology* 9:417-44. [Ten layouts of drawings.]

_________. 1983. Adaptive evolution within the Epipaleolithic of the Near East. In *Advances in world archaeology*, ed. F. Wendorf and A. Close, 2:99-160. New York: Academic Press.

Henry, D. O., B. H. Butler, and S. A. Hall. 1979. The late prehistoric human ecology of Birch Creek Valley, northeastern Oklahoma. *Plains Anthropologist* 24:207-38.

Henry, D. O., F. E. Kirby, A. B. Justen, and T. R. Hays. 1980. *The prehistory of Hog Creek: An archaeological investigation of Bosque and Coryell counties, Texas.* Tulsa: Laboratory of Archaeology, Department of Anthropology, University of Tulsa.

Marks, A. E. 1970. *Preceramic sites.* Publications of the Scandinavian Joint Expedition to Sudanese Nubia, vol. 2. Stockholm: Scandinavian University Books.

Marks, A. E., ed. 1976. *Prehistory and paleoenvironments in the Central Negev, Israel. Volume I: The Avdat/Aqev area, part 1.* Institute for the Study of Earth and Man, Reports of Investigations, 2. Dallas: Southern Methodist University Press.

_________, ed. 1977. *Prehistory and paleoenvironments in the Central Negev, Israel. Volume II: The Avdat/Aqev area, part 2 and the Har Harif.* Dallas: Department of Anthropology, Institute for the Study of Earth and Man, Southern Methodist University.

_________. 1983a. The Middle and Upper Paleolithic transition in the Levant. In *Advances in world archaeology*, ed. F. Wendorf and A. Close, 2:51-98. New York: Academic Press.

_________, ed. 1983b. *Prehistory and paleoenvironments in the Central Negev, Israel. Volume III: The Avdat/Aqev area, part 3.* Dallas: Department of Anthropology, Institute for the Study of Earth and Man, Southern Methodist University.

Marks, A. E., and H. L. Crew. 1972. Rosh Ein Mor, an open-air Mousterian site in the Central Negev, Israel. *Current Anthropology* 13:591-93.

Marks, A. E., and T. R. Scott. 1976. Abu Salem: Type site of the Harifian industry of the Southern Levant. *Journal of Field Archaeology* 13:43-60.

Marks, A. E., and P. W. Volkman. 1983. Changing core reduction strategies: A technological shift from the Middle to the Upper Paleolithic in the Southern Levant. In *The Mousterian legacy: Human biocultural change in the Upper Pleistocene*, ed. E. Trinkaus, 13-33. International Series 164. Oxford: British Archaeological Reports.

Nunley, J. P. 1971. Sociocultural units of the southwestern Texas Archaic: An analytic approach. Ph.D. dissertation, Southern Methodist University.

Schild, R., and F. Wendorf, eds. 1977. *The prehistory of Dakhla Oasis and adjacent desert.* Warsaw: Institute for the History of Material Culture, Polish Academy of Science. [Illustrated jointly with Iza Niewiadomska; drawings of bifaces predominate.]

Volkman, P. W., and D. Kaufman. 1983. A reassessment of the Emireh point as a possible type fossil for the technological shift from the Middle to the Upper Paleolithic in the Levant. In *The Mousterian legacy: Human biocultural change in the Upper Pleistocene*, ed. E. Trinkaus, 35-51. International Series 164. Oxford: British Archaeological Reports.

Wendorf, F., R. Said, and R. Schild. 1970. Egyptian prehistory: Some new concepts. *Science* 169:1161-71.

Wendorf, F., and R. Schild. 1976. *Prehistory of the Nile Valley.* New York: Academic Press.

_________. 1980. *Prehistory of the Eastern Sahara: Paleoenvironments and human exploitation.* New York: Academic Press.

Wendorf, F., and R. Schild, assemblers, and A. E. Close, ed. 1980. *Loaves and fishes: The prehistory of Wadi Kubbaniya.* Dallas: Department of Anthropology, Institute for the Study of Earth and Man, Southern Methodist University.

_________. 1984. *Cattlekeepers of the Eastern Sahara: The Neolithic of Bir Kiseiba.* Dallas: Department of Anthropology, Institute for the Study of Earth and Man, Southern Methodist University.

Wendorf, F., R. Schild, N. El Hadidi, A. Close, M. Kobusiewicz, H. Wieckowska, B. Issawi, and H. Haas. 1979. Use of barley in the Egyptian late Paleolithic. *Science* 205:1341-47.

Wendorf, F., R. Schild, R. Said, C. V. Haynes, A. Gautier, and M. Kobusiewicz. 1976. The prehistory of the Egyptian Sahara. *Science* 193:103-14.

Index

Numerals in boldface are figure numbers.

(continued)

(continued)